John Simpson Chisum

John S. Chisum

John Simpson Chisum

Jinglebob King
Of the Pecos

Mary Whatley Clarke

EAKIN PRESS
Austin, Texas

OTHER BOOKS BY THE AUTHOR

The Palo Pinto Story (Fort Worth, 1957)

(Editor) *Life in the Saddle*, by Frank Collinson; (Norman, 1963)

David G. Burnet (Austin, 1969)

Thomas J. Rusk: Soldier, Statesman, Jurist (Austin, 1971)

Chief Bowles and the Texas Cherokees (Norman, 1971)

The Swenson Saga and the SMS Ranches (Austin, 1976)

A Century of Cow Business (Fort Worth, 1976)

The Slaughter Ranches and Their Makers (Austin, 1979)

Library of Congress Cataloging in Publication Data

Clarke, Mary Whatley.
John Chisum: jinglebob king of the Pecos.

Bibliography: p.
Includes index.
1. Chisum, John Simpson, 1824–1884. 2. Ranch life — Pecos River Valley (N.M. and Tex.) — History. 3. Cattle trade — Pecos River Valley (N.M. and Tex.) — History. 4. Pecos River Valley (N.M. and Tex.) — Biography. 5. Ranchers — Pecos River Valley (N.M. and Tex.) — Biography. I. Title.
F392.P3C483 1984 976.4'9 84-18677
ISBN 0-89015-465-1

FIRST EDITION

Published in the United States of America
By Eakin Press, P.O. Box 23066, Austin, Texas 78735

ISBN 0-89015-465-1

With love to "Lash"
(Joe Brooks Whatley)
My favorite cowboy

Drawing of a jinglebob cow by noted Panhandle artist, Harold Bugbee of Clarendon.

Far up the Pecos in New Mexico, was John Chisum, most baronial, most feudalistic, in many ways most individual of all Texas cowmen that gave to the West its character.

J. Frank Dobie
A Vaquero of The Brush Country.

CONTENTS

PREFACE

I first learned about John Simpson Chisum and his adventurous life in the Territory of New Mexico in the 1950s when I was writing the "Cowboy Collinson" series of stories for *The Cattleman*. Collinson, an English cowboy, buffalo hunter and Indian fighter, worked on the Chisum ranch and knew Chisum, as well as Billy the Kid, and other lesser known, though colorful characters of that era. Collinson's stories about "The Cattle King of the Pecos," excited me and held my interest permanently, leading me to research Chisum's life and attempt this book about him some years later.

John Simpson Chisum was a trailblazer and lived at a time when red men and rustlers alike preyed on his herds and a man's life had little more value than that of a coyote. Though regularly confronted with danger, Chisum was not easily intimidated. He tried to avoid trouble, but if pressed would meet it head-on, or lead his cowboys into combat.

In writing about Chisum, Harwood F. Hinton, Jr. said, "Today we have only a blurred picture of a misrepresented and uninterpreted individual living in the shadow of a bygone era."

This author has tried to bring the figure and character of Chisum more fully into the "sunlight" in this biography, perhaps the first to be written about him. As the links of the historical chain have been joined, I have had the intense pleasure of seeing a real-life character emerge. To be sure, there are still patches in the Chisum outlines which remain in shadow, but enough can be seen to enable one to perceive a man in passionate conflict with himself, and others. There was nothing halfhearted about Chisum. An artist would have painted him in bold strokes and colors. He was strongly *good* and just as *bad,* a self-serving opportunist, who exploited the land and people of that day, who lusted after power, wealth and prestige, who wouldn't, in some cases, pay his debts, who permitted and probably ordered others to kill for him.

On the other hand, Chisum could be a man of his word, honest, loyal and generous to those he liked and respected. There is little doubt in this writer's mind that he loved the little slave girl he took to bed and board and grieved over her loss to the extent that he never married. He was a loving brother and uncle and would seemingly go to any length to help and be a true friend to those who treated him justly, and meant him well.

ACKNOWLEDGMENTS

"If research be the selection of bones of information from one stack to reassemble them in another mound," as Llerena B. Friend wrote in the introduction of her fine book, *Sam Houston the Great Designer*, then I acknowledge with gratitude the several authors whose "bones" I have picked in writing the biography of John Simpson Chisum. First and foremost are Maurice G. Fulton, author of *History of the Lincoln County War,* edited by Robert M. Mullin; William A. Keleher, author of *Violence in Lincoln County, 1869–1881*; Harwood F. Hinton, Jr., editor of *Arizona and the West*, who granted me permission to quote or paraphrase from his Master of Arts thesis, "John Simpson Chisum, 1877–84," Faculty of Political Science, Columbia University. Hinton had several personal interviews with William J. Chisum, John's nephew, during the fifties when writing his thesis. Will spent his youthful years on the Chisum ranch and told Hinton many things about his uncle, describing the Long House in detail, and so on. Hinton is a John Chisum enthusiast and continues to gather information about him, hoping to write a biography of this pioneer New Mexico ranchman. Many sources cited by Hinton in his thesis, later published in *New Mexico Historical Review*, helped this writer with her project.

Two additional books that perhaps tell more about pioneer Texas cowmen than any other published volumes, and which contributed greatly to my research and writing were *The Trail Drivers of Texas*, compiled and edited by J. Marvin Hunter, and *Historical and Biographical Record of the Cattle Industry and the Cattlemen of Texas and Adjacent Territory 1895*, by James Cox.

I am grateful to Paul Campbell, President of the Fort Worth Genealogical Society, for permission to use excerpts from the book, *From Ayr to Thurber*, by William Hunter McLean, which the society published in 1978. This book contains a journal "Conquering the West," by Mrs. David Hunter, who with her husband and sister-in-law spent two years, from 1876 to 1878, on the Chisum Ranch in New Mexico. Her journal, hitherto unpublished, and cherished by her family for nearly a half century, sheds new light on Chisum history.

I wish to thank Don C. King, Secretary-General Manager of Texas and Southwestern Cattle Raisers Association, for granting permission to quote or paraphrase from several articles that have been published in *The Cattleman* during past years as well as the use of several photographs. Some of the articles were written by this writer, including the series entitled "Cowboy Collinson." These articles were later put into the book *Life in The Saddle*, edited by this author and published by the Oklahoma University Press, Norman, Oklahoma. I am also grateful to this press for short quotes and paraphrasing from this book. Collinson knew John Chisum and the Chisum range, and his stories have added history and interest to this book. Another generous and helpful friend was J. Evetts Haley, who gave me permission to quote from his books.

Special appreciation is due Mary Charlotte Faris of Texas Christian University Library, Fort Worth, for her gracious help in locating relevant material. Sincere thanks go to the librarians of New Mexico who sent Chisum background as well as valuable books that were not otherwise available. The assistance of Arizona librarians was most helpful, and they too, have my gratitude.

Mrs. Eugie Jones Thomas and Mrs. Jinks Jones must be warmly thanked also for the collection of family photos and memories which they so generously shared with the author. Last, but not least, I thank my daughter, Mary Harper, for editorial assistance.

Chapter 1

The Famous Mark of John Simpson Chisum

Of the thousands of cattle that have grazed over the vast plains and valleys of the Southwest for the past century or more, only one herd was known by a mark instead of a brand. That mark was called the jinglebob and belonged to John Simpson Chisum, pioneer cattle baron of the Pecos. He was known, not surprisingly, as Jinglebob John, or The Cattle King of the Pecos.

The jinglebob mark split the animal's ear, causing one section to hang and bob, thus the name. The cut made the animal look as though it had four ears instead of two. There was nothing a rustler could do to change the jinglebob unless he cut off the animal's ears altogether or turned it into a muley. This often occurred and became a practice during those wild, untamed frontier days preceding and following the Lincoln

* [Editor's Note: The word jinglebob and/or Jingle-Bob is spelled one way in *The Cattleman* magazine and another in the *Handbook of Texas*. For easier reading, we have used it as a single word.]

County War, when rustlers assiduously plied their trade over the unfenced New Mexico Territory.

Chisum was most particular about the jinglebob mark. He wanted it performed exactly right. He hired men especially skilled with the knife to make the cut, as it required steady hands and a precise skill. When the animal's hair was long, it was often difficult to see a brand, but even a man with poor eyesight could see the jinglebob from a distance with ease. The jinglebob was achieved by splitting the ear at the top, then severing the two main leaders. The ear dropped back so that it hung toward the neck. Sometimes the jinglebob froze off in blizzard-like weather. Then it looked like a deep underlap.[1] Cattlemen were fascinated by the mark and their stories built it into a legend.

A cowboy on the Pecos in the 1880s, Jack Potter, described the correct way to cut the jinglebob mark as follows: "Catch the calf by the ear and stick the knife into the lower part next to the root — then start a short up-slope bringing the knife out on the top side of the ear and about halfway between the rest of the ear and the end. This will let two thirds of the ear drop down and swing around just like earrings." [2] After the ears were cut the steer looked both comical and frightening. One cowboy called it the devil's mark.

The jinglebob fascinated writers too. Wellman described the mark in his book, *The Trampling Herd*:

> Jinglebobbed, a steer had an appearance both ludicrous and repellent. With his sharp horns, six or more feet long, curved like scimitars and sharp as bayonets; his stupid half-frightened countenance; his lean, sway-backed body and long legs, four ears, two standing erect, and two hanging down like those of a hound, made him a very specter of an animal fit to frighten and amaze.[3]

One writer said it was difficult to describe the mark, that it would have to be seen to be understood. It was described as a long slit in the ear, a hide skinned off the shoulder and allowed to hang down, and one naive or highly imaginative writer, wrote that it was a piece of tin tied to the ear by a string with a paper folded inside describing the color of eyes, hair and date of birth of the animal and the name of the owner.[4]

T.E. Brown, Sr., Artesia, New Mexico, one of the authors of *Three Ranches West*, wrote the following explanation for the jinglebob earmark in a letter to the author July 10, 1978:

> You must take into consideration that this brand originated at a time when there were a large number of cattle running wild over the area that Chisum was ranching on. Many of those cattle had never been branded. He needed a distinctive mark so that an animal with that mark could be easily detected among other wild cattle.

One time in San Antonio, Texas, when attending the reunion of the Trail Drivers Association, Colonel Jack Potter and a group of friends were reminiscing about pioneer cow brands and marks. Someone mentioned the jinglebob but most of the group did not understand how the mark was made.

"Jump into my car and we'll go to the slaughter house, and I'll show you how the mark was made," Potter told them.

When they reached the slaughterhouse, a veal calf was roped and tied. Potter plunged his knife into the lower part of the ear next to the head, leaving about one-fourth of an inch for the bob to swing on, and then cut upward to the top of the ear. This left two-thirds of the ear to drop down. He pointed out that Chisum always made it a point when cutting the mark to show the cowboys how to avoid making the upper slope too long. The cut had to be made just so, to assure that it could not be altered to look like anything else.[5]

No one knows for certain how Chisum came up with such a mark. Early in his career the story was told that one day when ranching in Bolivar, Denton County, Texas, Chisum roped a calf, flipped it over and jerked up its head. Taking a small bowie knife from the sheath on his belt, he jabbed it through the calf's ear, telling his cowboys, "See, I started at the bottom center and ripped 'em straight out. Now one part stands up and the other bobs." [6] He quickly cut the calf's other ear in the same manner. Satisfied with the ear cutting, he said, "We'll call this spread the Jinglebob Outfit."[7] Thus was born the most unique and probably the most inhumane mark ever given any cow in southwestern cattle history and elsewhere in the world. It seems certain Chisum had given considerable thought to the selection of a mark that would set him and his animals apart from all other ranchers and their herds.

Chisum was one of the first cowmen to drive his herd from Texas into New Mexico Territory. He followed the historic trail blazed by two prominent Texas cowmen, Charles Goodnight and Oliver Loving. When Chisum died in 1884 Charles Goodnight said of him that he was one of the most distinctive characters that ever drove cattle over the Goodnight Trail from Texas.[8]

At one time Chisum was said to have more cattle and the largest spread of any ranchman in America. He claimed 200 miles up and down the Pecos Valley by right of discovery. In reality Chisum was a squatter for land that had not yet been opened to settlers. His far-reaching cattle empire stretched over one hundred miles along the Pecos River in the southwestern part of New Mexico Territory and his virgin range was grazed by 80,000 cattle, their ears cut with the well-known jinglebob. The cattle were wild and were mostly handled by men just as wild.[9]

The jinglebob mark was so well-known throughout the country that a stranger passing through on horseback knew he had reached the much-talked-about Chisum range. From the numbers of cattle grazing over miles and miles of prairie, each one looking double-eared, it was plain to see that Chisum's cowboys, better known as "Jinglebob Peelers," had done their work well.

Chisum's brand, the Long Rail, was also well known. It was a long, straight brand across the animal's side from the shoulder to the point where the hair ended on the flank. Later, because the Long Rail had been altered by rustlers and because Chisum transferred most of his cattle to a St. Louis

Commission firm, he felt it necessary to change to a different brand.[10]

Chisum used a copper branding iron to burn the rail believing that copper changed the color of the hair later. He instructed his cowboys to use the copper branding iron, believing it would offer protection against rustling. The brand did seemingly change the color of roan or light-colored cattle, but did not make any change in red or dark cattle.[11] Chisum cattle were always cut out of the roundups by the jinglebob mark and not the brand.[12]

Ralph Emerson Twitchell in *Leading Facts of New Mexico History*[13] wrote that Chisum, after driving his cattle west from Denton County, Texas, where he first ranched, used two brands. The first brand was a half circle. In 1865 he used a brand of two parallel bars. It is not exactly known just when he began to use the Long Rail. However none of his brands were as sensational or as radical as the jinglebob earmark. It was a good thing that members of the nation's first humane society, organized in New York in 1866,[14] never saw the Chisum cattle with their deformed ears. Had they done so a complaint might have been filed against Jinglebob John.

Despite the jinglebob mark Chisum continued to lose cattle. He learned that the herds of several cowmen he had fired the year before had miraculously increased. Chisum told his brothers, Pittser and James, that the increase was from his own herds.

"I cannot prosecute because it's impossible to prove title to an unbranded calf that has been weaned from its mother," he said.

After some thought Chisum devised a plan that he hoped might be helpful. He would burn a small decoy brand between the legs or on the flank. He talked it over with his brothers and they, too, thought it might work. As a result, Chisum drove several oversized bull mavericks from the east side of the river and after branding them as planned, turned them loose near a well-known watering place. There they could be easily seen.

"If someone puts his brand on them, and if the decoy brand is not seen, I will have proof that the cattle are mine," Chisum told his brothers.[15]

The plan was soon tested. Jesse Evans, a cowman en route from Feliz, New Mexico, first saw an "unbranded" steer near the watering hole. He was elated. How did such a fine animal escape the branding iron? Roping and tying the steer while his branding iron heated, he examined the hide for a hidden brand since the steer had not shed its winter hair. As the maverick struggled to get up, Evans saw the decoy brand on the inside of the flank. Grinning and nodding his head, he said, "jinglebob bait." He understood Chisum's ruse. Jinglebob John hoped to catch some hard-working cowboy entitled to any maverick caught on the public domain. Burning a question mark on the steer's belly he rode off, leaving the maverick to attract the next cowboy riding by.

The steer was rounded up several times and each time a cowboy took the bait he burned his own brand into the steer's hide. Bob Ollinger, a tough hombre said to have killed three men, rode by the watering hole one day and saw the innocent-looking steer. After roping the animal, he started to tie him when he saw the steer's belly and flanks were covered with brands. Cursing with disappointment, Ollinger branded a skull and crossbones into the maverick's side for good measure and jinglebobbed its ears.

"This will be a warning to the Jinglebob Outfit," he muttered. "This will be the last time a cowboy on a tired horse, runs you down and ropes you."

Bellowing with fear and pain the frightened maverick ran wildly toward the distant range. From that day to this, so the story goes, the ghost of the maverick, covered with various brands, roams the old Chisum range along the Pecos, its ghostly bellow echoing over the plains.[16] The legend is often told in bunkhouses throughout the Southwest when the TV is out of order and the cowboys have to entertain themselves.

Chapter 2

John Simpson Chisum Begins His Ranching Saga

Who was this man, John Simpson Chisum, whose personality was as original as his jinglebob earmark, whose thousands of cattle grazed a grassy empire that reached a hundred miles along the Pecos River in southwestern New Mexico Territory?

His name was sometimes confused with that of Jesse Chisholm of Chisholm Trail fame, but as far as anyone knows, the two men were unrelated.

Not knowing that John Simpson Chisum spelled his name differently from Jesse Chisholm, many old-timers gave "The Cattle King of the Pecos," all the credit for opening up the first cattle trail to Kansas. These old-timers argued that they were right, that they even had personal information to prove their case. One writer explained it this way. "Many honest old men had elastic memories."[1]

The spelling of Chisum's name could have originated when his father or grandfather enlisted in the War of 1812 and the name was misspelled on the roster. The ancient spelling was Chisholm — like that of Jesse Chisholm.

T.U. Taylor disputed the spelling of the name in an article headed "John Simpson Chisum and His Kin." [2] He wrote, "For several years I swallowed without investigation the statement that the name of the family before 1815 was Chisholm, but in the last year I began to trace it back by deeds, wills and actual signatures and I have found no documentary evidence that justified the statement and I can state here that it is my opinion, since the United States Declaration of Independence, the correct family name has been Chisum." [3]

According to cowmen who knew John Chisum personally, he was "burned and wrinkled by sun and wind." [4] He was not tall, was in fact rather small of stature, though well built. He had gray eyes and black hair and mustache.

Gus Gildea[5] described Chisum as medium-sized with shrewd eyes, his face sunburned to the color of leather.

"Chisum was one of the smartest men I have ever known," wrote J. Smith Lea, a contemporary and friend, "although he did not have that appearance and was rather inclined to make people believe that he was not so bright. He was never afraid of anything or anybody, and if he ever got mad, no one ever knew it, and he would never, to save his life, change a position he had once taken." [6]

Although Chisum had no wife, his ranch homes, wherever they chanced to be, were usually filled with family and friends. Many travelers passing through that part of the country, who had heard of Chisum and his hospitality, made it a point to stop overnight at his ranch to enjoy good food and conversation. His hospitable reputation was almost as well-known as his jinglebob mark. Chisum was reputedly a fair and impartial boss who got along well with his hands. Drifting cowboys could usually get work with Chisum, who treated them right if they did their job. It was said that Chisum never owned a pistol and that he knew both cattle and men.[7]

Many writers have written that Chisum never carried a gun, but his nephew, Will Chisum, emphatically denied this story. "I never saw him buckle a gun on, but he carried it in a holster, buckled to the right side of his saddle horn. I never saw him shoot at anything." According to the nephew, Chisum always kept a firearm, generally a single action Colt .45, within reach, whether riding a horse or in a buggy. Certainly he did not wear a revolver strapped around his waist, cowboy-like, but it is foolish to believe he went unarmed.[8]

One writer wrote that Chisum had a favorite saying about being armed. "A six-shooter will get you into more trouble than it will get you out of." His tactic was to win by persuasion or by appealing to self-interest. "He was adept at powers of sarcasm expressed in a friendly manner, then changing into derision accompanied by jabs of loud, jeering laughter. After such a verbal insult the majority capitulated and the matter was dropped." [9]

Chisum was a just and courageous man and did not make hasty decisions. He was a quiet and unpretentious man, simple in manners with no show of extravagance. "His firm-set mouth and strong lines of his face showed forth the power of his determination and purpose." [10]

He was born John Simpson Chisum in Madison County, Tennessee, August 14, 1824. Claiborne G. Chisum, his father, was born in Tennessee also, and was of Scottish ancestry. His grandfather, James, was a Tennessee State Senator from 1821 to 1823. Chisum's mother, Lucinda, was of German extraction. There were five Chisum children, including John, Nancy, James, Jefferson and Pittser M. The family immigrated west to Texas in 1837, settling in the area which later became Paris, Texas. That part of Texas was then known as Red River District.[11]

Although John had little schooling in his boyhood, he read and studied until he was fairly well educated.[12] As he grew older he became quite well-informed on current events and discussed them knowledgeably. He also wrote a clear hand and expressed himself well, as his business letters reflected.

John's father was one of the pioneer citizens of Red River

District. He built the first courthouse in the budding town of Paris. It was a brick building about fifty feet square, with a peaked roof coming to a point in the center. County officers were quartered on the first floor, while court sessions were held upstairs.[13] John Chisum, then a young man, probably helped his father build the new courthouse. *The Paris Daily Press*, September 16, 1878, reported that Ed Gibbons and John Chisum helped to construct the new courthouse.

By 1851 the town of Paris was a busy frontier settlement and was the county seat of the newly declared county of Lamar. John Chisum, who was a handsome and amiable young man, was well liked by the young people of the community. He was ambitious, but was undecided as to what course to follow. Until this time he had helped his father on the farm and in construction work. It was said that one of the young ladies of the village who may have been furthering her own ambitions, and with whom Chisum was keeping company, urged him to enter politics and run for county clerk. Liking and admiring the young lady, John succumbed to her pleas and announced for the office. Disappointingly, he was defeated by only a few votes, and soon afterward went back to work for his father for a couple of more years.[14]

As his political opponent's term in the county clerk's post drew to a close, John announced again for the office and this time was elected. However, John did not hold the office many months before he realized it was not what he wanted to do with his life. He loved the outdoors and felt convinced that his place was there rather than in a smoke-filled, airless office. Nevertheless, he administered the office conscientiously for a time, until desire for the outdoors grew too great and he resigned before his term expired.[15]

John's resignation apparently disappointed the young lass who had encouraged him so enthusiastically to run for office, and what might have become a romance was nipped in the bud, and the young couple drifted apart.

Although John did not know with any certainty what he would do after leaving office, it was in the back of his mind to pursue the cattle business, convinced that it would become the leading industry of Texas. Cattle were the main topic of conversation wherever men gathered and most settlers had small herds. John's father had cattle, and so did their neighbors. Some day there would be a market for them and John wanted to be in on the game.

Then, suddenly, the totally unexpected happened. A man named S.K. Fowler, a New Yorker with money to invest in cattle, arrived in Paris. He soon met John Chisum and took a liking to the young man. As a result a partnership was formed between the two. Fowler would invest $6,000 in cattle and John would buy and look after the animals.[16] Chisum left immediately for Colorado County in the southern part of Texas and bought stock cattle at six dollars per head, the calves thrown in, and drove the herd to Denton County. He also bought one hundred yearlings at $2.00 per head, delivered.[17] There he established a ranch where the present-day Bolivar, Texas, is located. Although Chisum eventually owned over 200 acres of land around his ranch headquarters, he did not own an acre of range land.[18]

One wonders why Chisum drove his cattle to Denton County until one remembers how often he had heard his father describe that verdant section of Texas where the elder Chisum had ridden and chased Indians in 1841. The senior Chisum was a member of the company of Colonel Edward R. Tarrant, for whom Tarrant County, Texas, is named. The company had been organized to fight marauding Indians.[19]

Captain John B. Denton, also a member of that company, was killed in the Battle of Village Creek six miles east of Fort Worth on May 22, 1841. His body was brought back on horseback and buried in an unmarked grave on the west bank of Oliver Creek in Denton County. Nearby a tree was marked to show the location of the grave.[20]

In 1861, John Chisum, then ranching at Bolivar and remembering the story his father had related about Captain Denton's death and subsequent burial on the creek bank, searched for and found the old grave. Chisum brought the remains to his ranch home and buried them in a box in the corner of his yard on Clear Creek. In 1901, the Pioneer Association of Denton County removed the remains and buried them with appropriate ceremonies in the southeast corner of the Denton County courthouse lawn. The town and county of Denton are named for this Texas pioneer. John Chisum's father witnessed the first burial of Captain Denton and twenty years later the son witnessed the second burial.[21]

John Chisum's ranch home, built in 1854 on the crest of a hill, looked down and outward on a panoramic view that prompted one pioneer visitor to remark that a casual observer could spot a redskin a half mile away from the front porch. Chisum bought two more small tracts of land and altogether owned 220 acres, but his cattle grazed over the open, unfenced country for miles in most directions. The herd thrived and multiplied and the 1860 census valued Chisum's part of the livestock at $50,000.[22]

Water was hauled from a nearby creek during the first year of Chisum's ranching career at Bolivar. Then one day a traveling well digger came through the country and stopped at the ranch. He hoped to make his way to the California gold fields but was short of cash. Chisum made a deal with him. He would finance his California trip if the well digger would first dig him a well. The trade was made and the Bolivar ranchman later saw the well digger happily on his way West as the former sipped cool, pure water from his new well. That old well was one of the first ones drilled in Denton County and was used for many years. Today it is abandoned and covered with logs. This writer and her brother visited the old Chisum ranch in April, 1979, and saw the pioneer ranch home and abandoned well.

Although Chisum spent most of his time on the Bolivar Ranch he often rode into Denton and visited friends there. He was one of the charter members of the Independent Order of Odd Fellows Denton Lodge No. 82, IOOF, which was organized in 1859.[23]

Chisum's Denton County ranch home was known for its

hospitality and many good times were enjoyed there. Often an all night dance was held. John, who could make a fiddle talk, often joined the musicians playing for the dances, but made certain not to miss his turn dancing with the local and out-of-town belles. It was said that one time after a heavy rain, the dancers could not go home due to swollen streams. The rain continued until the creeks ran over their banks and the guests stayed on. Chisum had to kill an extra beef to feed them. One old-timer later recalled that three dances were held each day during the entire week, the longest dance party ever held in North Texas.[24]

John Chisum had another talent besides playing the fiddle. He was an excellent cobbler and made and mended his own boots and those of his cowboys. He had all the tools necessary and plenty of bark-tanned leather. Laces, when needed, were usually made from squirrel hides.

Chisum did not fight in the Civil War but was designated as a beef supplier to troops in the Trans-Mississippi Department, helping to supply the Confederates with beef. Cattle were gathered from several counties by Chisum and his cowboys, and driven down the Red River Valley to Shreveport where they were sold at $40 a head.[25]

Chisum's own ranch suffered when he was away on those long trail trips. He wrote James Waide, a neighbor, a discouraging letter in March 1862, expressing the uncertainty and disappointment of ranching:

> I got back from Vicksburg a few days since. I find the prairie all burnt off and we have had no rain hear [*sic*] since last spring. All my hands are gone, all my horses are gone. I am left behind in charge of 6 other stocks besides own.[26]

Chisum did not keep the Confederate money for any length of time that he was paid for his cattle. Instead he enlarged his profits by turning the money into cattle at range prices as fast as possible. When the Confederacy fell, he held little worthless money.[27]

By 1862, Denton County was rapidly absorbing new settlers and the free range was dwindling. Chisum realized he would have to move his herds farther west where there was still an abundance of space and grass. His ten-year-partnership deal with Fowler was drawing to a close and he decided to go it alone. Chisum told Fowler that he would either sell his own interest to him, or buy Fowler's interest. He told his old partner he planned to run only Chisum cattle in the future. Fowler, who lived in New York, sold his interests to Chisum in return for the latter's Denton County land.[28]

Before leaving Denton County, Chisum said goodbye to some of his family, including Jensie, a comely mulatto slave girl who had been a part of his life since his ranching days at Bolivar. Liaison between Chisum and Jensie came about in this way: late one evening a white family enroute to California stopped their tired team in front of Chisum's ranch home. A man, his wife and several of their children, as well as a young slave girl, were in the wagon. They camped near the Chisum home that night and the man visited with and told Chisum that he needed cash to continue the journey. He offered to sell the young slave girl, Jensie, who was fifteen years old, to Chisum for $1400. "She says her pa was white, and she's a good worker, keeps a house better than most whites," the man said, adding, "she's one of our family and I hope to buy her back one day."

In need of a cook and housekeeper, Chisum bought Jensie and soon realized what a treasure she was. She was an excellent cook and kept an immaculate house. She was easygoing and pleasant. She was young and pretty. The two became lovers and two daughters were born of that union in the next few years. The oldest daughter was named Harriet, the youngest, Meady. Chisum took Jensie to Gainesville, Texas, to have her babies delivered by a physician, and before driving his herds west he established the girls and their mother in a Bonham, Texas home, leaving money to care for them. At one time he visited them and left $100 for Harriet's schooling. It is doubtful that he ever saw them again after settling in New Mexico. One wonders what thoughts and feelings

Chisum had when telling Jensie and the two little girls good-bye. Was there regret and sadness in his heart?

Harriet, the older daughter, married at an early age and moved to a different part of the state and the family has never been able to reestablish contact with her.

Meady grew to womanhood and married John Dalford (Bob) Jones, son of a white Arkansas farmer and a slave mother. Bob and Meady settled in the Medlin community of Tarrant County, and reared a large family. Jones accumulated rich farmland which he farmed and ranched for a living.

Two of Bob and Meady's children, grandchildren of John Chisum, were living when the author wrote this book and she had the pleasure of visiting in their homes and hearing their story from them. They were petite Mrs. Eugie Jones Thomas in her nineties and Jinks Jones in his eighties. Another brother, Emory, died in 1968. All families lived in the Grapevine Lake area of Tarrant County. Much of the land that their father owned is now under the lake. Jinks and Emory, who were partners, operated an auction barn for many years near Grapevine, and currently Jinks operates a weekly flea market at the same location. The brother and sister cherish enlarged photographs of their grandfather and spoke with pride of their heritage.[29]

Chapter 3

Chisum Moves Herds to Concho River

At the close of the Civil War Chisum fulfilled his contract to furnish beef to the Confederates and no more trail trips were made to Louisiana and other southern points. Although other cowmen were beginning to trail outfits north, it was said that Chisum never drove his cattle in that direction. He did drive two small herds to Little Rock, Arkansas, after the war, selling them to a small packinghouse in which he had an interest. Chisum sold most of his cattle from his own home range, choosing not to seek a special market. Cattle buyers often came to his Bolivar Ranch and bought entire herds of cattle at a time.[1]

Western fever was spreading. Since camp followers from the Civil War were coming into Texas, many making their living by stealing and slaughtering cattle for their hides, and because the Indians were continuously on the rampage, cattlemen began moving their herds west both for more range and protection. It was safer to live on the plains where marauding Indians and rustlers could be seen before an attack. New settlers also were moving into Denton County and filing claims on land. Chisum, watching his own range begin to dwindle, soon caught the western fever himself, and in 1863 drove his first herd of 15,000 cattle to a new frontier. Most of his Denton County cowboys went with him.[2]

Arriving on the Concho in Southwest Texas, Chisum set up a temporary camp in a grove of pecan trees about two miles up river from the junction of the Concho and Colorado Rivers. In what would be known later as Coleman County, Chisum and his men built log huts for living quarters and log pens for corrals. The nearest neighbor was twenty-five miles away.[3]

Some of Chisum's cowboys returned to Denton County to drive back more cattle and others stayed on the Concho to establish the new ranch headquarters. When the Civil War ended most of Chisum's cattle had been moved west to the new range. Each drive was made through Indian country and the cowboys were constantly in danger from redskins in ambush. All men were well armed and prepared for attacks on the western drive. The cattle were no doubt driven through Tarrant, Parker and Palo Pinto Counties, a route well populated by hostile Indians. Night watching was particularly dangerous, especially in the full of the moon, and all hands slept with their guns. The trail trip from Bolivar to the Concho covered approximately 240 miles and took about three weeks to travel.[4]

Since Chisum's cattle were the first large herds to be driven into Coleman County, the animals banqueted off the entire open prairie to the southern part of the county. Luckily, there was a log trading post a few miles distant which was owned by Bill Franks, a jovial chap who had felled timbers from the river to build his trading post, the first in the county. He sold groceries, some ranch supplies and whiskey. John Chisum, who was thankful supplies could be bought in his new homeland, became Franks's best customer. Franks got a kick out of teasing and playing jokes on the cowboys and, as a result, his store was called "Trick-em." Later, when applications were made for the first post office in Coleman County, Trick-em was among them. It was accepted, but the spelling was changed to Trickham.[5] Chisum himself later bought the store at Trickham, operating it for several years as an adjunct to his expanding ranching business, thus saving time and expense of sending his hands after supplies to Fort Worth.[6]

Chisum hired Emery Peters to run the store for him, but sold out to L.L. Shield in the 1870s.[7] His herds grazed over the southern part of Coleman County and his jinglebob earmark and Long Rail brand were well known. Brands were not recorded in Coleman County until 1866, therefore Chisum's brand is not counted among Coleman County pioneer brands.[8]

After four years in Coleman County, 1863 to 1867, Chisum was again faced with the same problems that pushed him westward from Denton County.[9] Daily, caravans of covered wagons could be seen rolling across the west and southwest plains of Texas and Chisum could not hide from himself the fact that free grasslands would soon be a memory. Chisum listened to increasing talk extolling the virtues of the New Mexico Territory farther west. New trails were opened and cow buyers and drovers began to drive cattle to the military forts and Indian reservations in New Mexico Territory.

Chisum was acquainted with Charles Goodnight and Oliver Loving who drove the first cattle from their Palo Pinto County ranches westward over the old Butterfield Trail from Ben Ficklin in Tom Green County to Fort Sumner, New Mexico, where they sold their herds. The community of Ben Ficklin, the first county seat of Tom Green County, was named for a Confederate soldier and organizer of a local pony express who received a government mail contract and established the Concho Mail Station, later to be called Ben Ficklin. When flood destroyed the village of Ben Ficklin in 1882, San Angelo was named the county seat. At one time, Tom Green County was called "The Mother of the Conchos," because a dozen other counties were eventually carved out of it.[10]

Goodnight drove a herd up the east side of the Pecos

River in 1866 and afterward described to Chisum the width and grandeur of the virgin range on the west side of the river. Curiously, large herds of buffalo which had completely devastated the east side of the river, had never crossed over to graze the west side.

John Chisum sent his brother, Pittser, who had followed him from Denton County to the Concho Ranch, on an inspection trip to the Pecos. Chisum instructed Pittser to ride over the territory carefully so as to bring back a realistic assessment of the quality and scope of the cattle range. Pittser spent months dutifully carrying out his brother's orders and returned with glowing reports. He told John not a hoof of stock was seen from Fort Sumner all the way down to Horsehead Crossing, Texas. Spurred by Pittser's report, Chisum decided to look over the new land for himself before moving there permanently. He wanted to drive a herd with him, hoping to sell the animals to Fort Sumner where beef was at a premium.[11]

In 1867, Chisum and his cowboys headed west for the Pecos with a herd of 900 head of cattle. Chisum had no worry about losing the trail as long as he could follow the South Concho, but beyond that, according to Goodnight, it was "rough and very rough," while Horsehead Crossing over the Pecos was the cowman's nemesis.[12]

An ideal spot for Indian attacks, Horsehead Crossing was also utilized as a camping site for roving bands of rustlers. After leaving the South Concho, Chisum's herd had to cross ninety miles of desert without water before reaching the river. Chisum, who was accustomed to hardships, felt that if others had made it, he could too, and set out on the drive hoping for the best.

Horsehead Crossing is at the site where the Pecos River crosses part of the boundary between Pecos and Crane Counties. The place got its name from the horse and mule skeletons found on both sides of the river. Horse skulls were found there as early as 1850 by R. Bartlett who camped at the crossing while on a surveying tour of the Mexican boundary.[13]

Thousands of cattle died of thirst near Castle Gap before

reaching Horsehead Crossing. A pioneer cowman named Campbell lost 2100 head near Castle Gap. His cowboys, suffering from thirst, took the wagon and headed for the Pecos for water. During their absence, the mule Campbell was riding died of thirst and Campbell himself almost died before his cowboys returned with water. At about this time, S.R. Coggin, who had left his ranch in Brown County, began to pass aged steers headed east toward the Concho. When he reached Castle Gap, Coggin saw Campbell, who told him that his herd was lost and he was financially ruined. Only a few hundred head survived the formidable ninety miles over the desert country. Coggin drove the remaining cows with his herd to Bosque Grande where he sold them to Chisum, mindful to send Campbell his share of the money.[14]

Collinson, the English cowboy and author of the book *Life In The Saddle*, worked for a time on the John Slaughter ranch in Arizona. Slaughter told him about losing a herd at Horsehead Crossing. Having started to Arizona with a Texas herd, Slaughter went up the Nueces Canyon expecting to find water. None was located. Another 100-mile waterless drive was made. On the third day the cowboys drove the suffering cattle within six miles of the Pecos where they smelled water. It was night. The river banks were high and steep. Slaughter, who felt it was best to wait till daylight to attempt watering the cattle, ordered the hands to guard the animals till morning. He had not, however, paid enough attention to the wind, which was high and strong. At midnight, there was a change and the cattle, now downwind, could smell the water. Wild with thirst, they stampeded toward the river in a mad run, only to fall to their death from the high bluffs. In daylight, it was sickening to see 2,000 cattle, piled dead, one on top of the other, in the river. But Slaughter, like Chisum and others of the same mold, was a stoic who did not give up easily. He returned to Texas and drove back a second herd.[15]

The Pecos River heads in the distant snow-covered mountains of northern New Mexico and the pure waters from melted snow gradually turn red and brackish as the alkaline plains are reached. The river winds its sinuous course to the southern plains through desolate greasewood country into the Texas canyons. Charles Goodnight, who cut out the first trail west, told Evetts Haley, who wrote a book about him, that he had counted thirteen graves at Horsehead Crossing when first crossing the river there. "All the result of pistol shots but one," he said, "and I shall never forget the impression made upon me by these lonely graves where rested cowboys killed in battle with one another after having fallen out while crossing the long stretch without water," he said.[16]

Lieutenant French, an early-day army officer wrote the following description of the Pecos River:[17]

> The Pecos is a remarkable stream. Its waters are turbid and bitter, and carry both mechanical mixture and chemical solution, more impurities than perhaps any other river in the south. Its banks are steep, and, in a course of 240 miles there are but a few places where an animal can approach them for water in safety. Not a tree or bush marks the course; and one may stand on its banks and not know that the stream is near. The only inhabitants of its waters are catfish; and the antelope and wolf alone visit its dreary, silent and desolate shores. It is avoided even by the Indians. (Reconnaissances, Lieutenant French)

Chisum never forgot his first cattle drive into New Mexico. His herd made it without mishap up the South Concho past the painted rocks near Paint Rock and over the prairie to Ben Ficklin, on and on they plodded through the Needles (sentinels and pilots of the Concho Country), hugging the South Concho until every drop of water disappeared.[18] From there they began the ninety miles from the South Concho headwaters to Horsehead Crossing. The cowboys faced the burning, westward sun many wearing, monotonous days before arriving at Castle Gap. From there, they could see Horsehead Crossing in the hazy distance.[19]

Before Horsehead Crossing was reached on Chisum's drive, the cattle were suffering for water, their tongues swollen, their eyes sunken in their skulls. A few miles from the crossing they smelled water and as one cowboy said, "All Hell broke loose." The cattle started to run. Chisum shouted, "Turn them north, they smell water," as he tried to head off the bawling, suffering animals. But the cattle could not be stopped. "They won't ever make it to Sumner," one of the cowboys dolefully predicted, but strangely enough they did.[20]

T.U. Taylor of Austin went over the old trail followed by Chisum in 1936 from Bolivar, Texas, to Bosque Grande, New Mexico, a distance of some 700 miles. The drive through the Indian country of Denton, Tarrant, Parker, and Palo Pinto Counties would not compare with the harrowing drive across the ninety-mile desert country from the South Concho to Horsehead.[21]

Profiting from his first drive, Chisum learned several valuable lessons. Although Goodnight lost hundreds of cattle at Horsehead, he continued to cross his herds there and recrossed them near old Seven Rivers to the east side to avoid west side tributaries. Choosing not to follow Goodnight's example, Chisum drove his herds up the west side of the Pecos and braved the dangerous tributaries so his cattle could have fresh water. Thus two trails from Texas into New Mexico were created, Goodnight on the east, Chisum to the west. Immigrants traveling by covered wagon, followed the Chisum Trail as they moved into New Mexico in the seventies. The Chisum Trail on the west was also followed by the stage that ran from Carlsbad [then Eddy] to Roswell in the 1890s. The first railroad was also built over the old Chisum Trail in 1894 from Carlsbad up the Pecos to Roswell.[22]

Chapter 4

Chisum Makes a Deal With Goodnight

Holding on to horses was one of John Chisum's most demanding tasks. Before he moved from the Concho country to the Pecos, Indians swooped down one night and drove away all his horses, making it necessary for him to ride to San Antonio to buy more. Since horses were stolen at every opportunity by roving bands of Indians, they were at a premium. Most cattle drives sustained thefts of horses. Horses on consignment were sometimes stolen before they could reach the buyer. For instance, James Daniel and his crew of cowboys were sent to receive a herd of horses for Chisum. When camped at Dagger's Bend on the Pecos, all horses were stolen by the Indians, including the cowboys' mounts. They had to walk several hundred miles before finding any horses for sale.[1]

Comanches and Apaches, hating the intruding settlers, plied their vengeful raids between Red River and the Pecos, stealing horses and stampeding and killing cattle. In July, 1874, a group of Indians rode boldly to the Chisum store at Bosque Grande, where a remuda of 125 picked saddle horses was being held, and stampeded and drove them off. That same month, Indians drove off forty more head of horses from a camp set up by Pittser Chisum near Twelve Mile Bend, and on the same day attacked a roundup crew under McKittrick and drove off more horses. That fall still more horses, about 160, used by a branding crew, were driven off at Eighteen-Mile Bend. Two line riders, Newt Huggins and Jack Holt, were killed in that raid. Chisum was so short of horses during 1874–1875 that little branding could be done when 13,000 calves were dropped. Much of the cow work was done on foot until new mounts could be bought.[2]

In 1874, Indian troubles were so prevalent that Chisum realized he would have to take personal action to stop the raids. Since the military troops gave no protection, he took matters into his own hands. Rounding up and arming one hundred men, he marched them near the Mescalero Indian Reservation, went alone to the officers' headquarters and paid for drinks for the house. After getting the officers drunk, Chisum and his armed band attacked the Indian village about one mile from the post and killed 175. After that, the Indians did little stealing from the jinglebob herds and feared any cowboy who worked on the Chisum range.[3]

Chisum maintained several cow camps in New Mexico Territory after he first moved into that region and settled at Bosque Grande. There were Chisum camps at Black River and Three Rivers, with a large camp at Fort Stanton. Bosque Grande, which served as Chisum's permanent headquarters for several years, was a large cottonwood grove on the river below Fort Sumner. Chisum regarded it as an ideal headquarters site.[4] He eventually bought the store and residence owned by James Patterson and by 1873 was running 80,000 cattle, was furnishing beef to the Apache Indians, and employing around one hundred cowboys. He told friends he could fill an order for 40,000 cattle at a short notice. He laid claim to the rangelands for several hundred miles along the Pecos — from Anton Chico on the north to Seven Rivers on the south, and from the White Mountains on the west to the Canadian River on the east. His cattle deals were as far flung as his range. At that time, Chisum was without doubt the best known cowman in America and controlled the largest land holdings.[5]

Although there was actually no law in the vast territory and rustling was continuous, Chisum decided to take his chances with the handful of other pioneers who had staked claims on that virgin land. He liked the country. The mountain peaks, a far cry from the prairie country where he had ranched on the Concho, seemed to keep watch over his cattle kingdom. He found the mountains, El Capitan, the Guadalupe and Sacramentos constant inspirations.

Lincoln County contained 27,000 square miles and was the largest county in America at that time.

It was unfortunate that this huge New Mexico county adjoined the arid belt of western Texas which was a natural sanctuary for desperados and dishonest cowboys. Trouble usually resulted from their visit to the outside world. Murder was no stranger to Lincoln County. Bands of desperate characters lived off the honest citizens of the county, plundering or hiring out as mercenaries of one feuding faction or another, paid by the day to fight battles.[6]

There is no doubt that Chisum, too, hired guns from the same ranks during times when Indian raids and rustling were rampant. If the men did their work he was not concerned about their morals.

Rustling was so bad that Chisum later hired a group of cowboys commonly called "warriors," experienced gunmen familiar with six-shooters and Winchesters, who helped him patrol his far-flung domain. Since Lincoln County joined the arid belt of western Texas from which cowboys and desperados swarmed, it was easy to hire warriors to guard his ranch who were familiar with all types of firearms.

His warriors caused a temporary time of peace on the range. However, *The Independent*, July 28, 1877, complained that things were disgustingly quiet, that Chisum would no doubt drive out as many as 10,000 cattle during the year and mark and brand 20,000 calves. Three thousand head had been delivered to Colonel R.D. Hunter, some of the cattle were on the trail to Kansas at that time and the remainder would be driven to Arizona.

The peace did not last very long since there were not sufficient civil authorities to hunt down the bands of thieves working along the Ruidosa, Penasco and Hondo Rivers. *The Independent*, September 8, 1877, reported:

> For a long time our country has been infested with a band of thieves, desperados and outlaws, who have openly defied the laws — Their enmity appears to have been directed against our largest tax payers, evidently for the purpose of alarming them into buying their safety . . . When some uncommon atrocity would arouse public sentiment against them they would flee to Donã Ana County to rusticate, generally taking with them a drove of horses or herd of cattle from our citizens.

Times were so precarious on the open range that travelers through the country were in constant danger from holdups and other violence. On one occasion, according to *The Independent*, October 6, 1877, the mail carrier who had left for Silver City by muleback came upon a bunch of Chisum men camped by the road. Getting an "Indian scare" he ran his mule down and returned to town on foot bringing the mail with him.

There were Mexican cattle thieves too, who would steal cattle at night and run them across the White Sands to the Rio Grande ranches. Negro Frank, Chisum's faithful helper, said, "Such thefts kept Mistah Chisum short-handed."

Chisum realized he could not stop all the rustling on his range since it was too big and the cattle too numerous to be patrolled too closely. His warriors could only do the best they could but even they could not stay up night and day to guard his cattle. He was not too rough on a rustler when one was caught. Once a man named Lucero was caught stealing Chisum cattle. He was well flogged by Chisum's men and warned if he tried it again he would be killed.[7]

Bad luck plagued Chisum with the first herd he drove from the Concho into New Mexico. He soon learned that reverses were part of the hazards of the trade. Although he made arrangements to sell his first herd to James Patterson who had a contract with a New York firm to furnish beef to the Fort Sumner Indians, he never received a penny for the cattle. He had been paid for the herd with drafts amounting to $23,000 but the paper was worthless as the firm had gone bankrupt.[8]

Before returning to Texas, Chisum was contacted by Charles Goodnight, the veteran cowman who had driven some of the first herds west. Goodnight proposed that Chisum help him fill a contract for beef for Fort Sumner. At that time, Goodnight was living many hundreds of miles distant in Pueblo, Colorado.

Goodnight, a native of Illinois, moved to Palo Pinto County, Texas in 1857. He served as an Indian scout and ranger during the pre-Civil War days and during the war acted as both guide and scout for a frontier regiment. By 1865 he had accumulated a nice herd of cattle, but had no market. Oliver Loving, commonly known as "the dean of Trail drivers," who moved to Palo Pinto County in 1855, had successfully driven a herd to Chicago in 1858, and later to Denver in 1859 by way of the Arkansas River and Pueblo.[9] He and Goodnight, who were neighbors and friends, combined their herds in 1866 and Loving piloted the route of their first successful cattle drive to the West. The trail they cut became the historic Goodnight–Loving Trail which was later followed by thousands of cattle.[10]

Goodnight and Loving drove their herd over the Butterfield Overland Mail Route from Fort Belknap, Young County, Texas, by way of the middle Concho and Castle Gap to Horsehead Crossing. The cowmen left the mail route at that point and drove the cattle up the Pecos, crossing and recrossing the river as the terrain and need for water required.[11]

On the second drive in 1867, the same year that Chisum was driving his herds to the Pecos from the Concho, Loving and Bill Wilson rode ahead to see about selling the cattle at Fort Sumner. They were attacked by Indians on the Pecos and Loving was critically injured in the side and wrist. He sent Wilson back to warn Goodnight. Loving was found by

Mexicans and was taken to Fort Sumner where post physicians diagnosed blood poisoning and amputated one arm. However, it was too late to save Loving's life.

Goodnight arrived at his bedside before Loving died and promised his old partner that if he did not recover he would bring his body back to Texas for burial. Goodnight also vowed to continue the partnership until the current deal was closed. Three weeks later, Loving died. His body was temporarily buried in a large box of charcoal for preservation, and, in the fall, was taken by wagon to Weatherford, Texas, for burial.[12] Cowboys served as bodyguards. G.A. Holland described the funeral procession in his book *History of Parker County and the Double Log Cabin*:

> Slowly the funeral cortege covered the 600-mile trail between Fort Sumner and Weatherford — passing as an apparition through New Mexico, across the desert, then northeast over brush country and plains to its destination. With bowed head and aching heart, Goodnight, the great cattleman who throughout his life played a stellar role as leader and comrade, rode alone on horseback behind the improvised hearse, which carried all that was left of his beloved friend.[13]

Chisum agreed to deliver 7000 head of cattle to Goodnight that first year. He planned to round up the cattle in Texas and drive them to Horsehead Crossing. There he would be met by his brother, Pittser, and his crew of cowboys who would drive the herd up the Pecos to Chisum's headquarters at Bosque Grande. At that point the herd would be turned over to Goodnight, who would take the cattle westward to Arizona and Colorado for sale to the government forts and Indian reservations. Chisum returned to Texas where he gathered up 1165 good steers, paying twenty dollars in gold for each, and delivered the animals to Pittser at Horsehead Crossing. He then returned to the Concho for more cattle.

Before starting up the Pecos to Bosque Grande, Pittser asked an army officer and his soldiers who were camped at the crossing to escort his herd to its destination because he feared Indians might try to stampede the cows. The officer refused and Pittser and his crew had to drive the animals themselves. They made it safely as far as Black River in the southern part of New Mexico where they camped for the night. During the night, however, as Pittser had feared, Indians stampeded the herd and not a single cow was recovered. For years afterward, the bones of the animals were seen strewn throughout the Guadalupe Mountains where the Indians had apparently left them after slaughtering the cattle for meat.[14]

John Chisum, who had just returned to Horsehead Crossing with another Texas herd, learned the bad news from his brother. Forcing a smile, he said, "Well, here's another herd, try her again." [15]

After the Indians had stolen and driven off the entire herd under Pittser's care, Chisum went back to Texas to gather up more cattle to fulfill the Goodnight contract. Most of the cattle Chisum eventually drove back had individual brands. This caused Goodnight to believe they were stolen property. He refused to accept them.[16] Chisum only laughed good-naturedly and held no ill will toward his partner. He had no difficulty disposing of the herd elsewhere for cows were at a premium during those frontier days. But those Texas brands would return to haunt him later in a court of law.

Altogether, Chisum delivered cattle to Goodnight for about three years. The two men had an agreement that Goodnight would pay Chisum one dollar per head profit over and above current market prices in Texas, due to the risk Chisum took driving the cattle into the territory. During his dealings with Goodnight, Chisum delivered approximately 15,000 or 16,000 head of cattle and profits were divided equally. Goodnight said that Chisum used his part of the money to increase his herd which was said to reach some 60,000 head in its heyday.[17]

Later, when Chisum drove his own herds to Colorado and Arizona the trails were beaten out and could be easily followed. Chisum drove the herds west along the Hondo for sixty miles, then southwest along the Ruidoso, and up through the White Mountains. He skirted the White Sands on the south until he reached the Organ Mountains. From there, he followed the southern slope to San Augustine Pass, and down the western side of the pass to the vast plain of the Rio Grande. He crossed the river where Las Cruces is located today, then headed west through Cook's Canyon, past Stein's Peak into Arizona.[18]

Chisum had no close neighbors when he settled at Bosque Grande, but R.K. Wylie drove a herd of 9000 cattle into the Pecos soon afterwards and camped above Chisum's place. That winter, a blinding snowstorm struck the range and Wylie's cattle drifted south, mixing with Chisum's. The morning after the storm Wylie climbed out of his dugout and studied the range through his field glasses. He did not see a single animal. Mounting his horse, he rode to Chisum's place and told him about his loss. Chisum laughed heartily and told Wylie not to worry. "How much do you value your lost herd?" he asked. Wylie replied that the cattle were worth $65,000. A deal was soon made and Chisum said the cattle would be rounded up in the spring with his own herd, that he had contracted for all the cattle he could sell. He would pay Wylie at that time. Wylie rode back to his camp a much happier cowman.[19] It was rumored that although Chisum had bought the Wylie herd he thought it best for some reason to let the cattle remain in Wylie's name.[20]

Wylie and Chisum remained good friends and at times Wylie worked for Chisum and was in charge of one or more of his cow camps. They had several partnership deals. Wylie came to Texas from Mississippi in 1847. He received his first cattle in payment for work on a brick chimney. He worked for various ranchmen, drove cattle to Old Mexico and made a profit of $500. Upon returning to Texas, Wylie met James Patterson on the Concho and ranched with him there and along the Pecos. Wylie also ranched near Ballinger, Texas, for several years, but returned to the Pecos in 1876 with Sam R. Coggin. At that time, they received 8000 head of cattle from Chisum and drove them into Texas, establishing the Matador Ranch.[21]

During Chisum's first flourishing years on the Pecos his

major worry was not being able to dispose of his stock as fast as it increased. Such an item was published in the *Grant County Herald*, Silver City, April 11, 1876:

> We hear of cotton being king, of railroad kings, but J.S. Chisum of Bosque Grande is our stock king of New Mexico. We remember upon one occasion, when finding our stock king in a deep reverie, of asking him why he was in such deep study. "I'm in great trouble because I cannot dispose of my stock as fast as it increases," he said.

A reporter from the *Las Vegas Gazette*, visited Chisum's ranch November 15, 1875, and wrote the following description:

> The Chisum ranch is situated on the Pecos River, New Mexico. It extends south along the river near Fort Sumner to Seven Rivers, a distance of 150 miles. This is the extent of the ranch north and south. East and west it reaches as far as a man can travel on a good horse during a summer, practically as far as stock can range from the river without water. The country on both sides of the river is high rolling prairie, covered with a thick, heavy growth of black grass, the most nutritious of grasses. The grass is knee high on every hill and mesa this year. The river is the second in size in the territory and is fed from the snows and springs of the western mountains. To the eastward extends the Staked Plains from which side there is not a single tributary to the Pecos. The home ranch is at Bosque Grande, fifty miles below Fort Sumner. At this point a general store is kept, buildings and large corrals constructed. At convenient points up and down the river from the home ranch are secondary stations, or cow camps, which the herds occupy as temporary homes. Not less than 100 employees take care of the cattle. The expenses of Mr. Chisum, for horses the last ten years have been very heavy, due to losses from Indian raids. At one time, after an Indian foray, only ten head of horses were left on the ranch. The savages captured at that time three or four hundred head. The men had to herd on foot until other stock could be purchased. The horses have been of valuable quality, ranging from a dozen to 1000 head. This year, however, has been unusually quiet. The Indians have not committed a single depredation. The spring roundup will take place in February.[22]

Chapter 5

John Chisum Makes Two Major Decisions

John Chisum made two major decisions in 1875. He decided first to move his ranch headquarters from Bosque Grande forty or fifty miles to South Spring River where a gushing stream of pure water intrigued him. There he could, in time, have shade trees, a garden, orchard and flowers, and with proper irrigation, fields of grain. He traded 2400 head of cattle to James Patterson for forty acres of that rich river land only five miles from the budding town of Roswell. For a time he lived in a square five-room adobe house which opened onto a small patio built by former owners. Chisum would later raze the structure, build a fort-like adobe home of primitive elegance and call it the Long House.[1]

Then, with stunning effect, he made his second decision, amazing the cattle world. After engaging in ranching some thirty years, he was transferring the majority of his stock holdings to Hunter, Evans and Company, the well-known St. Louis, Missouri, commission firm.[2] News of the sale spread like a fire over the territory and into neighboring states. It was unbelievable that the Cattle King of the Pecos was selling out.

The *Pueblo Colorado Chieftain*, December 3, carried the following item:

> The ranch of John S. Chisum of Bosque Grande, New Mexico, was sold to Colonel R.D. Hunter of St. Louis the other day for $215,000, one half down. Colonel Hunter — will engage more extensively than Mr. Chisum in the breeding business.

The Mesilla News, Mesilla, New Mexico, reported that Colonel Hunter, who bought one of John Chisum's ranches for $312,940, had 221 fine bulls from Kentucky to place on his new spread.[3]

Colonel Hunter, who handled cattle deals for his firm, was nationally recognized as one of the best known commission men in America at that time. As he matured, the West attracted him. He successfully mined for gold in California in 1849, and after the Civil War (in which he served on the northern side) he led cattle drives. With the colorful Shanghai Pierce of Texas cattle fame, Hunter established a shipping point at Ellsworth, Kansas, during the trail-driving era. He also successfully supplied oxen to cross-country wagon freighters and even drove Longhorn herds up the trail from the Texas coastal country.[4]

Hunter and Albert C. Evans established their commission firm in 1874 and in 1876 were headquartered in East St. Louis with branches in Kansas City, Chicago, Kansas and various Texas locations.[5]

A quote in *Parsons Memorial and Historical Library Magazine* said this about their firm:

> Among cattle commission firms none is more notable as being composed of substantial, practical, clear-headed business men than that of Hunter and Evans and Co. Each member of the firm is a successful livestock man of long experience, which, coupled with their individual responsibility, renders their house one reliable and safe.[6]

Of course Chisum knew Hunter's reputation as did most

ranchers throughout the southwest. About 1873 or 1874 Chisum borrowed a large sum of money from the company through Colonel Hunter. Hunter agreed to accept cattle for the loan.[7] Then, in 1875, as stated earlier, Chisum transferred most of his stock holdings to Hunter and Evans. The exact terms of the acquisition are not known, or whether the negotiations were verbal or documented. It is known, however, that Chisum stipulated that $27,000 worth of stock be turned over to his brother, Pittser, when the acquisition was consummated, so as to "defraud, hinder and delay" settlement to his creditors who were unable from that time until Chisum's death to collect because he had no assets they could trace back to him.[8]

Pittser also was permitted, as the deliveries were made to Hunter and Evans, to select several thousand heifers as payment for the ten years he had worked for his brother, John. When the bulk of the consignment had been filled in 1879, Pittser received title to the herd.[9]

Chisum had always appreciated good bulls and had bought as many as he could find through the years, buying some of his first bulls from Goodnight.[10] He paid from $100 to $1,000 each for them. Since Pete Maxwell was a neighbor and a pioneer cowman in the area, Chisum bought as many highgrade bulls from him as he could spare. The result was a good Shorthorn herd before Herefords were driven into New Mexico.[11]

Chisum discarded the Long Rail brand after transferring his vast herds to Hunter and Evans. He needed a new and individual brand for the smaller graded herd he had kept. Early in 1876 he learned that his brand had been altered into a pitchfork, a Lazy F attached to a rail, and a pigpen. He was shown these unrecorded brands in the slaughter pen at Fort Stanton where the cattle had been sold. It was easily seen that the Long Rail had been altered. After seeing the pigpen brand he said, "When they get to using my rail to build a pigpen, it is time for me to squeal."[12]

Chisum did squeal "openly and vainly," threatening Lawrence G. Murphy whose range joined his own in the

Seven Rivers country, but Murphy only laughed. Where was the proof? Recourse to the law was a mockery as Murphy controlled the courts.[13]

Murphy's cattle ranch was a mysterious adjunct to his Lincoln store and saloon. His herds were small, never averaging more than a few thousand head at a time, yet he sold thousands of steers annually. "Either his cows were endowed with amazing fecundity, or else . . ." [14]

In his government report Frank W. Angel later wrote:

> L.G. Murphy and Company had the monopoly of all business in the county — they controlled government contracts and used their power to oppress and grind out all they could from the farmers and those who were opposed to them were forced to leave the country.[15]

Chisum's ambition to build up a good smaller herd was another reason why he made the deal with Hunter and Evans. He could see ahead to a time when he could no longer run such large herds. The free range was gradually being claimed by newcomers. New families had already settled on some of the land he had originally claimed. He did not own the land, but claimed it by "right of discovery" since it was unsettled, virgin territory when he arrived.[16]

Was there still another reason why Chisum transferred his herds and most of his property to Hunter and Evans? Did an underlying fear urge him to do so? Why did he want Wylie to keep the cattle he had sold to him in his own name?

About 1876, Hunter and Evans held a business meeting with Chisum. It was time to settle some of the debt Chisum owed them. It was decided that Hunter would send men and equipment to the Pecos to work with Chisum in rounding up and cutting out the cattle. When the herds were ready, Hunter and Evans crews would drive them off the Chisum range.[17]

Colonel Hunter sent his younger brother, David, an experienced cowman, to the Pecos to supervise the cattle roundup on the Chisum range. As part of the contract to help build up the herd, David Hunter would drive a herd of thoroughbred cattle consisting of twenty heifers from one to three years old, over one hundred bulls from one to two years old, and one full-grown bull to the Chisum ranch, a distance of 450 miles from St. Louis.[18]*

The trail trip began the last of September or first of October 1876. Six men and a cook drove the cattle and the grown bull was led all the way by a ring in its nose. Margaret Hunter, David's wife, and his sister, Lizzie, went along on the trip, traveling in an ambulance, as some early-day conveyances were known. Besides the cattle there was a remuda of some thirty or forty horses, four mules for the chuck wagon and two mules for the ambulance.[19]

Weeks of travel and adventure followed which Mrs. Hunter never forgot. But let Mrs. Hunter tell the story in her own words:

> I shall never forget that first night. Along about two o'clock in the morning we were aroused from a restless sleep by such terrible yells and howling. Lizzie and I clung together with fear. I cried, "Dave! Dave! Come here! Unfasten the curtains. Tell us, is it Indians? Oh hurry!" My husband came, but only laughed at our fears and said it was only coyotes that would not harm us.

*New information regarding the Chisum–Hunter and Evans deal, published for the first time, has come to light in a new book entitled *From Ayr to Thurber. Three Hunter Brothers and the Winning of the West.* The author is William Hunter McLean, a prominent Fort Worth citizen, and a descendant of the R.D. Hunter family. The book was published by the Fort Worth Genealogical Society in 1978 and includes a journal written by Mrs. Margaret L. Hunter, wife of David Hunter, who rode west from St. Louis to the Chisum ranch with her husband. She called her journal "Conquering the West." It contains many interesting bits of New Mexico history that were never published until McLean's recent book came out.

Margaret Hunter's story continues:

> The first day, and for many more days after, we traveled over prairie country, sometimes on a road or a trail, and part of the time trying to find a road and water for the horses. We had a big keg of drinking water with us. Over this buffalo country

where the buffalo hunters had killed thousands for the hides, the buffalo bones were scattered white over the prairie.

We arrived in Dodge City in a downpour of rain, glad to get under shelter and to be where it was warm — There was much talk among the cattlemen about the fool "hoe men" that were fast settling on the Kansas prairies.

Dodge City was noted as the toughest town in the West. It was an outfitting center for buffalo hunters within a radius of hundreds of miles, and that year, 1876, she became the great cattle center of the West. About this time the Atchison Topeka and Santa Fe Railroad was being built west from Dodge City. This filled the town with bums who follow up new railroads. Fort Dodge, a government post, was located only five miles distant. The wild and woolly cowboys were celebrating in the "Lone Star" dance hall. The hall was jammed full of buffalo hunters, cowboys and street girls, drinking and fighting, and through the night some would go galloping, yelling and shooting off pistols. We had our horses in a corral, but it was hard to corral our boys —

We drove ahead to locate campsites, where we could find water and fuel. When the boys first spoke of picking up chips I looked about and saw no trees and asked, "Where can you find chips? There are no trees." The boys smiled as they picked up some gunny sacks and said, "They are buffalo chips."

There was great excitement when we saw our first herd of buffalo. They were grazing a few miles to the southwest of us. We stopped when within a few hundred yards of them. At first they seemed to pay no attention to the boys, but when the shooting started, the animals dashed off on a mad run. A big bull came loping toward the ambulance wagon. I ran to hold the wagon mules, screaming to Lizzie for my gun; but the cook, or one of the boys was using it. We and the mules were badly frightened, but the buffalo was turned from us. Several animals were killed, but only the tender hump loins were dressed for food. That night we camped where we had a wood fire and broiled buffalo meat held over the fire on a sharpened stick. My, how good the supper tasted!

. . . We had crossed and traveled along many streams without fear. The one that left its imprint on my memory was Red River. It was bank full, a raging torrent. The men held council to decide what was best to do and finally decided to stack every-

> thing in the mess wagon up as high as possible, and we did the same in the ambulance. It was a difficult matter to get the horses to take to the water, hard to keep them from drifting with the current down the stream. The horses and mess wagon were taken over first, then two of the cowboys fastened their lariats to the ambulance wheels on the side up stream, one his rope to the ambulance tongue, then the mules and horses started to swim. The boys swam their horses up stream to prevent drifting with the current. We sat with our feet up on the seat, held our breaths and held on tight. It was rather a bad crossing . . . When we reached the foothills, mountains or bluffs of the Staked Plains [no doubt the cap rock] we had a hard pull to reach the top. The mules had to be whipped and coaxed to make them pull. I felt the ambulance would have been tipped over backwards but for the lariat rope hitched to the tongue and the man on his good horse helping pull us up. I had often wondered why cowboys wore such high heeled boots and spurs, but they had demonstrated the need of them on this trip.[20]

The Hunters finally reached Bosque Grande on the Chisum range. At that time there was a well-built five-room adobe house on the site. The house had two bedrooms, a large kitchen and a long dining table flanked by benches. There was a large storeroom in the back where the cowboys bunked. The deep glass windows were shuttered and the two outside doors were barred at night.[21]

The Hunters did not sleep in the house but camped in the cottonwood grove. The following morning Margaret and Lizzie went for a walk, looked at the muddy brown water of the Pecos and gazed in disbelief at the strange-appearing jingle-bobbed cattle grazing over the far-reaching range. They agreed that it would take a skilled surgeon to cut the ears so uniformly.[22]

The following day they drove down to South Spring River to John Chisum's headquarters and were amazed by their host's fine, spacious home, but soon learned it was a fitting extension of "the kind, gracious man who welcomed them so warmly." [23]

Chisum made his guests feel at home at once. He had two

comfortable rooms ready for them. There were no womenfolk on the ranch at that time, only John Chisum, his brother Pittser who was somewhat smaller of stature, and McKittrick, who ran the store. Pittser looked after the ladies, taking time to show them around and being considerate of their welfare. The winter was mild so the women often rode horseback into Roswell for the mail and made friends in the village. David Hunter and his crew worked long hours with Chisum and his cowboys rounding up cattle.[24]

Chapter 6

"Judge Lynch" Takes Over

When the cowboys working for Chisum meted out justice for murder, theft, and other crimes, it was said facetiously that "Judge Lynch," the court of the range, took over.[1]

John Chisum was a fair man, but he could be punitive when he believed punishment was due. According to a story told by Frank Collinson in his book, *Life in the Saddle*,[2] Charles Goodnight and S. R. Coggin arrived at the Chisum Ranch one night on their way to buy cattle. A young wrangler penned their horses for them. The next morning when the men went for their horses, another cowboy called the young wrangler an insulting name for no apparent reason. "I won't take that," the young wrangler responded.

Angered, the challenger declared, "You'll take this," and fatally shot the youth. As a crowd of sympathetic cowboys crowded around the victim, the killer pointed to the dead wrangler, saying, "There's the trouble."

Since the slain boy was a kind of favorite of most of the cowboys, the gunslinger was ordered to drop his pistol while the men tied his hands. They planned to hang him after breakfast. Realizing that the men were serious, Coggin said to Chisum, "Stop them, John, or they'll hang that man."

"What can I do about it?" Chisum replied. "Fall branding time is here and I need every hand that I've got. It would take an entire week to go to Vegas and bring back the law and then all of us would have to stay in town to be witnesses at the trial." Looking Coggin straight in the eye, Chisum concluded, "The cowboys on the ranch are just as good judges as could be found in Vegas. I'll stay out of it. He killed the boy, why not hang him?"

After tying one end of the rope to the limb of a cottonwood tree and the other end around the killer's neck, the cowboys made the man mount his horse. Then they led the horse from under the man they had condemned and left him swinging in the air. Next morning when the roundup crew left for work, Chisum commented, "We are short two men, but we'll make out." [3]

A story was printed in the *Kansas City Livestock Indicator*, March 7, 1889, which Chisum had told a reporter several years earlier. According to Chisum, he and his cowboys chased a horse thief who had stolen three mounts from the remuda and caught him the following day. Chisum was quoted as saying:

> We asked him no questions. Vegetation was scant there, but we took the highest we could find and dragged him up until his head was within two inches of the limb — the buttons of his clothes gave way, and when we left him he was almost as naked as when he was born.

Another time all the cowboys on the Chisum ranch struck for higher wages, excepting Beaver Smith, the cook. Chisum paid them off, telling them he needed a rest and that they would get no more food at the ranch due to the fact that they had gone on strike.

Angry at Beaver for not striking, the cowboys decided to brand him. "Let's put the Rail on his ribs, then he'll belong to Chisum the rest of his life," they said. Throwing the beleaguered cook to the ground, the men burned the Rail on his ribs. They even attempted to cut the jinglebob on one ear but failed, leaving it scarred and bloody.

Frank Collinson, the English cowboy, who later worked on the Chisum ranch, became acquainted with Beaver. The cook showed Collinson the brand on his ribs. Collinson further said Beaver told him that after the cowboys had branded him, he told them to get their horses and beds and get out for he would kill them if they stayed on the Pecos. When Beaver died in Las Cruces, the Rail brand was still visible on his ribs.[4]

One night on the trail, a herder abandoned his guard several hours before his watch was over and rode into town. Chisum, who was in camp and on his pallet, saw what happened and realized the stock would soon begin to drift, or worse still, stampede. He fired the man as soon as he had the opportunity, demonstrating again that he was not a man with much tolerance for insubordination.[5]

Chisum hired hundreds of cowboys during his ranching career. Most of them were loyal. They knew they would be treated right if they gave Chisum an honest day's work. The cowboys were a Duke's mixture of both good and bad. The majority drank hard when they had a chance. For instance, when some of the cowboys were driving cattle to Arizona in 1877 to fulfill a Chisum contract, they stopped at Dowlin's Mill in Tularosa County where whiskey was sold. The boys bought a round of drinks. After a bit, Johnnie Evers began playing with his gun and shot himself in the leg. He had to be taken to Fort Stanton for treatment. Another cowboy, Ramon Garcia, shot and killed J.M. Franklin, one of Chisum's most dependable men. A hothead when in his cups, Garcia was arrested, but was killed en route to Fort Stanton when he tried to escape.[6]

Gus Gildea, a cowboy on the Chisum Ranch in 1876, relates some of his adventures on the cattle trail in the popular book, *Trail Drivers of Texas*.[7] Gus helped round up cattle for two Chisum herds. A mixed herd would be trailed to Nebraska and another herd, composed of old "Moss-horns," would be delivered to the Indian reservation at San Carlos, Arizona. At that time, Chisum had the contract to furnish beef for the government for some 7000 Indians. Gus helped drive the herd of 4000 "Moss-horns" along with 150 horses to Arizona. The route customarily followed passed through the Mescalero Indian Reservation. Chisum usually gave the chief twenty head of cattle as he trailed through with his outfit. This time, however, the Indians were camped outside the reservation. They were in the trees along the Tularosa and Lost Rivers on the west side of the White Mountains where they had no right to demand toll. Gus and the cowboys banded together against the Indians, refusing to give them any cattle. When Chisum rode forward to meet Solomon, the Indian chief, Gus asked his boss to turn back and not interfere, threatening to shoot the chief if he did. Chisum rode back to the wagon, whereupon Gus rode up to the chief. Pointing in the direction of the reservation, he told Solomon, "Vallese, No bakeshee." (Go, no tip.) It is interesting to ponder on that. Where did Gus, a simple New World cowboy, hear the expression "Bakesheesh," or "Bakshish," which for hundreds of years has rung familiarly in streets and alleys of Middle Eastern cities, called out by beggars enlisting alms from passerbys. Doubtless the word "Vallese" is corrupted Spanish, deriving from *Vayase,* the Spanish word for Go, in the command sense, i.e., Be off with you. It is also interesting that the Indian understood him.

Though angry, the chief whirled his mustang paint and, trailed by his warriors, made off down the trail at a fast gait. When the lead cattle neared the reservation Gus and the cowboys rode ahead to make sure the Indians would not stampede the cattle. There was no trouble. Evidently the Indians had not forgotten the raid on their reservation the year before by Chisum and his cowboys who killed some 175 warriors (Chapter 4). Chisum said that was the first time he had not paid the Indians with at least twenty head of cattle for trailing through the reservation.

That was a lucky drive for Chisum. After they had

watered the cattle at Lost River they had to go sixty more miles over desert to San Augustine Springs, a drive of some thirty-six hours. Chisum usually lost cattle from exhaustion or thirst on this leg of the journey, but not a head was lost, Gus reported.[8]

When the Rio Grande was reached, the cattle were permitted to graze and Chisum sent some of his men to Doña Ana to pick up stolen cattle. Gus was appointed "Side Boss" on those scouting trips, taking four cowboys with him. By noon, they had gathered fifty head, each wearing the Chisum brands.

The Mexican farmers who used the animals as work oxen, followed the herd trying to retrieve the cows, but Gus and his cowboys resisted their efforts. The men dismounted and found protection in a deep ditch where they fired volley after volley of shots at the Mexicans, driving them back and killing seven. An officer from the fort, who had been told that cattle thieves had taken the oxen, dropped his investigation into the incident when he learned that the cattle indisputably bore the Rail brand and had jinglebob earmarks. Without doubt, the animals had been stolen from Chisum and driven from his range.[9]

Gus wrote that "Judge Lynch," an apparent reference to Chisum and/or his hired men, tried and executed two men at Bosque Grande for murder during the roundup. Since murder had been committed in cold blood in both cases and witnessed by the cowboys, the murderers were tried, found guilty and executed by the cowboys. Afterward, they went about their work as usual, Gus related, since part of the code of the law on the range was to "forget it." Further discussions might have led to more trouble.

Gus quit the Chisum outfit when they reached San Simon in Arizona as there remained only thirty-five more miles to Croton Springs in Sulphur Springs Valley where the herd would be delivered. He did, however, work for Chisum again in 1878 in Sulphur Springs, guarding his range north from Pinal Mountain south to present-day Pierce, Arizona, and from the Dragoon Mountains east to the Chiricahua Mountains.[10]

Chisum was often seen on Old Sandy, his favorite horse, with his binoculars about his neck. He dressed like one of his own cowboys in a broad brimmed felt hat, a shirt of flannel without a vest and with his trousers stuffed into his boots. It was said that he once turned down a million dollar offer for his ranch while attired in a twenty-five cent straw hat, a sixty-five cent shirt, seventy-five cent overalls and twelve-dollar brogans. He said he would rather look at a cow than a dollar.[11]

Chisum was away from home often due to business deals, frequently traveling by stage. In 1876 while en route from Silver City to Mesilla with his attorney, the stagecoach was attacked by three masked bandits when it rumbled into Cook's Canyon. The driver was ordered to stop and get down from his seat, while passengers were told to line up outside the coach. Chisum had $1,000 in currency in his pocket but unseen by the bandits, he stuffed it into his boot. When ordered to turn over his money, he handed the highwaymen a handful of change. The bandits got away with several silver bars which were being sent from Silver City to New York.[12]

J. Frank Dobie tells the following story about Chisum in his book, *Cow People*. Whether factual or legendary it is worthy of repetition. On another of his trips John Chisum stopped at the Brown Place Hotel in Denver. All the waiters were black and were dressed in white in the spacious dining room where meals were served in courses. "No soup," Chisum stated as a waiter hovered over his chair with the tureen. Soon another waiter, dressed in white, stopped at Chisum's chair.

"Soup, sir?" he enquired.

"I told you I did not want soup," Chisum replied, more firmly. When a third waiter appeared and attempted to ladle him some soup, Chisum's voice rose in unmistakeable anger.

"How many times do I have to tell you that I don't want any soup?"

Before his stay was over, Chisum must have wondered if he would not have been better off sleeping on his bedroll under a tree on the outskirts of town. The room he was assigned to at the hotel had been used for weeks by a "lunger," the term used to identify a consumptive in those days. Each night around midnight, an attendant dressed in white, had been in the habit of coming into the sick man's room to give him an injection. Chisum did not, of course, know this until it was too late to do anything about it.

Meanwhile, he spent a lively evening in the hotel, playing poker and enjoying the conviviality of other companions before finally going off to bed. He was almost asleep when an apparition in white glided into his room, leaned over and injected a hypodermic needle into his arm. Groggy with sleep and no doubt the liquor he had consumed earlier, Chisum cried out, "Well, damn your soul — you are getting that soup into me after all." [13]

Chisum kept a close eye on the marking and branding of his stock to make certain both were done as he desired. When he saw a mistake, Chisum would chide the cowboy about it good humoredly, but insist on a speedy correction. There was little about cow work that Chisum did not know.[14] His cowboys respected his opinion. Charles Goodnight stressed that no one had any advantage over Chisum as an old-fashioned cowman, that he was the best counter he ever saw. He said Chisum could simultaneously count three grades of cattle going at a trot with striking accuracy.[15]

Chisum was quick to spot someone else's mistake, too, as was demonstrated in one cattle cutting exercise. It was essential at such times to see that a calf followed its mother. Evetts Haley, Texan and historian, recalled that incident in his book, *George W. Littlefield, Texan*, whose historic brand LFD was almost as well known in Pecos Valley as that of Chisum's. Haley wrote that one day at the roundup Chisum saw a freshly branded LFD calf suckling one of his U cows. Phelps White, Littlefield foreman, rode up and Chisum called his attention to the calf, saying, "Mr. White, I wish you could have been around here while ago and heard that little calf talking to his mama. 'Mama! mama!' he said, 'those old LFD boys just chased me round and round, roped me and dragged me, and put that old LFD on my side, Mama, when I wanted a nice little U on my shoulder like you.'"

"Now," Chisum said, "if you put another LFD on him, it would just spoil the poor little thing's hide. Just let him go Mr. White, but I wish you'd be a little careful hereafter." [16]

Chisum had "cow sense" and this trait came to his and his neighbor's aid many times. During the 1870s when trailing his Texas stock to the Pecos he learned that a herd behind him wasn't making much progress due to stampedes. Chisum rode back to investigate, accompanied by Sug Robertson, a cattle inspector, who was riding with him at the time. Chisum asked permission to ride through the herd and soon located the trouble. He told Robertson it was a steer with wide and crooked horns, with only one eye. "Mr. Chisum ordered that the steer be cut out and driven down the river and shot," Robertson said, "and there was no more stampeding." [17]

Frank Collinson praised "Nigger Frank" as a faithful ranch hand, who had been with Chisum since slave days. Chisum had purchased Frank for $700. Frank revered Chisum. He told Collinson that when he was a little boy Chisum would ask him to sleep at the foot of his pallet to keep his feet warm on cold nights.[18]

John Chisum loved music and in his younger days fiddled for pleasure or amusement. He often played with the musicians at the dances held in his ranch homes. He owned a good violin. One of the Pecos cowboys was named Bell Hudson. He was also a musician. It delighted Chisum to listen to Bell play for him in the evenings and he insisted that the cowboy use Chisum's good violin instead of the makeshift instrument Bell owned.

When Bell Hudson started to work for Chisum, other cowboys were being hired at the same time, and Chisum wrote their names in a small black notebook which he carried in his pocket. He asked Bell if he wanted to be known by his real name or a nickname. Bell replied that he wanted to be known by his real name and was grateful for this choice later after one of the cowboys, known by the nickname of Judge, was drowned in a swollen river and no one, not even Chisum, knew his real name.

Pecos cowboys always sang the following verse after riding in after a long, wearing trail trip:

When I get home, I'll sleep a week,
And all these vows I make I'll keep.
I swear by rope and bridle rein,
I'll never drive jinglebob cows again.[19]

Chapter 7

The Chisum War on the Pecos

As cattle thefts continued, Chisum worried about the loss of stock belonging to Hunter and Evans. Their agreement with Chisum specified that all cattle would remain on Chisum's range until Hunter and Evans directed certain herds trailed out. Chisum feared that rustlers might think he would not guard his range as carefully since the cattle no longer belonged to him. If anything, he was more conscientious and redoubled his efforts to protect the cattle he must gather up and deliver to Hunter and Evans.

The increased cattle rustling gave rise to suspicion and distrust, eventually turning neighbor against neighbor and evolving into a bitter cattle feud that would gradually become known as the Lincoln County War which would rage over the virgin Territory of New Mexico for the next several years. In October, 1876, a cowboy named Yopp, who was in charge of the Wylie camp near Seven Rivers, complained to his helpers, Buck Powell and Dick Smith, that too many cattle were disappearing. A quarrel resulted in which Yopp was killed.

Ash Upton, of Roswell, a newspaperman, sent the following to a New York newspaper:

> At Wylie's camp, 80 miles below here, about six weeks since, a man named Yopp in charge of the herd became enraged at an employee, Buck Powell. He drew his revolver and fired three shots at him. Buck seized a Winchester, and shot Yopp in the mouth. Yopp fell and remained for a few minutes insensible. Suddenly recovering, he re-opened fire on Powell. Powell's gun hung fire. He then seized Yopp's own gun and shot him through the heart. Powell did not receive a scratch. He wanted to go 150 miles and give himself up for trial, but was persuaded not to. So the matter ended.[1]

The killing of Yopp was the beginning of the "War on the Pecos" or "Chisum's War," which was a forerunner of the Lincoln County War.

In a separate incident Chisum learned that suspicious activities were also occurring around the Beckwith corrals at Seven Rivers. As a result he sent his brother Pittser, Jim Highsaw and other cowboys to investigate. They found several hundred ears of jinglebob cattle buried near the corrals.[2] As far as Chisum was concerned this was proof that his cattle had been stolen, driven off and sold.

Chisum and Highsaw rode into El Paso where they recovered some of the cattle. More important, they got descriptions of the men who sold the animals. When they returned home, Highsaw went to Smith, a sometime Wylie foreman, and accused him of stealing Chisum's cattle. A heated argument ensued in which Highsaw beat Smith to the draw and shot him.[3]

Though this was the second death on Chisum's range, the latter merely remarked that conditions were "looking bilious" and took no immediate action. Real trouble was brewing, however, and Chisum knew it. He soon paid a visit to Fort Stanton and discussed the problem with Colonel George F. Furrington. The Colonel refused to send out the cavalry for private protection, so Chisum went to Sheriff William G. Brady for help. He got nothing for his trouble. Brady's excuse was that Seven Rivers, where so much trouble occurred, was in Doña Ana County and out of his territory.[4]

After Smith's death, some twenty of his friends rode to Chisum's South Spring Ranch determined "to get" Highsaw, who was laying low in the fort-like ranch complex. Chisum's other men took cover behind the parapet of the roof, pointing their rifles through portholes at the angry gang on the ground. After delivering themselves of a few threats and curses, however, the men rode away without firing a shot.[5]

The Beckwith Ranch was rumored to be the stronghold of a gang of cattle rustlers. When they learned stolen jinglebob cattle had been driven to the Beckwith Ranch, Chisum and his men made up their minds to clean up the situation. They headed north up the Pecos toward the ranch, thirty strong and well-armed men, determined to mete out justice in cowman fashion.

Upon reaching the ranch headquarters, Chisum's men drove off every horse and mule they could see and cut off the water supply to the house. They held the Beckwith ranch home under siege and Chisum sent a note asking that the women and children vacate the premises. The women refused, saying they would rather stay where they were than entrust themselves to Chisum and his men. On the third day of the siege Chisum ordered an attack on the walls of the house, but the initial enthusiasm of his men was wearing thin. They refused to follow Chisum's orders, saying they did not want to fight a war. Then it was learned that Buck Powell and Charles Woolsey had escaped during the night and were riding into Mesilla for help from the law. Chisum realized the game was up and one of his men, under a flag of truce, went to talk with William A. Johnson, a son-in-law of Beckwith. Johnson said a truce would not be possible until Chisum and Wylie paid the money they owed to certain ex-Chisum cowboys in the Beckwith ranch house. Chisum and his men returned the livestock and turned the water back on, then headed south for the spring roundups. Chisum made no settlement with the cowboys over old debts at that time.[6]

It was several weeks later that Deputy Sheriff Andrew Boyle of Seven Rivers rode into the Wylie camp with warrants for the arrest of Chisum, Wylie and Highsaw. They found Chisum confined to the camp with smallpox, too sick to be disturbed. Boyle and his deputies rode on to the South Spring Ranch where Wylie and the other hands were entrenched in Chisum's fort-like house. No fight took place and the sheriff's posse rode off. Chisum recovered from the dread smallpox but, still weak and sickly, it was a while before he returned home.[7]

Fulton wrote that Andy Boyle was as ruthless a fellow as could be found in the country, and seemingly became involved in every mischief that might be afoot.[8] Boyle, who was associated with the Catron-Murphy-Dolan interests, later prepared his report of the Beckwith attack. The long, rambling epistle was published in the first issue of the *Mesilla Independent*, June 15, 1877. Fulton wrote that Boyle's report could not be ignored, although in some respects it was one-sided.[9] The report follows:

Mesilla, New Mexico
Doña Ana County, June 15, [18] '77

T.B. Catron. U.S. Attorney:

Sir: In accordance with your request I give you the facts of the Chisum war on the Pecos River. John S. Chisum claims to have about 40,000 head of cattle grazing on the Pecos River; also R.K. Wylie has about 3000 head. Those two stock men concluded they wanted all the Pecos River to themselves, a distance of 200 miles.

1st. A man by the name of Richard Smith, Wylie's foreman, was killed by a herder, James M. Highsaw, on March 28, in camp, shot in the back five times with a Colt's improved 45 cal. six shooter. Chisum remarking that the war had commenced and he had six more men to kill (the Seven Rivers thieves) viz. Lewis Paxton, Underwood, Charles W. Woltz, R.W. Beckwith, W.H. Johnson and Buck Powell, these men having each from 200 to 1000 head of cattle which they were holding in the neighborhood of Seven Rivers. Lewis Paxton happened to go to Lincoln

City on business on April 8, when he was informed by Major Brady, Sheriff of Lincoln County, that John S. Chisum had been hiring men to kill him and others. Mr. Paxton got on his horse and rode home as fast as possible to let the men know what Chisum intended doing. These men then left their camps and came up to Seven Rivers to the house of H. W. Beckwith for self-defense.

2nd. On April 10, James J. Dolan came down to Seven Rivers after beef cattle which he wanted for the use of the U.S. Troops at Fort Stanton. A Party of these men started to go down to Paxton's and Underwood's camp to get the cattle when they met six of Chisum's and Wylie's men on the public road near Wylie's cow camp, who, seeing this party coming down the road, got off their horses and got down into a water course and prepared to shoot at them, which Mr. Underwood's party seeing, charged down onto them. They fired a few shots on both sides — nobody hurt. Chisum's men then got on their horses and rode into their camps, remarking that they had got whipped. Chisum then rode off to Fort Stanton to try to get soldiers and arms to assist him in killing these men. The commanding officer told Chisum it was none of his business, he was not there to kill citizens but to protect them. Chisum then tried to get the sheriff of Lincoln County to assist him, he also would not do it.

3rd. On April 20, Chisum and Wylie got all the men that would join them together, some thirty men, went to Seven Rivers, took what horses and mules those other men had grazing, went up above Mr. Beckwith's house about a mile and cut off the water in the ditch that supplied the house, saying he would starve them out, and when they should have to come out to get water, they could waylay them and shoot them down. That night, two men, Chas. W. Woltz and Buck Powell, rode past them and went to Mesilla, the county seat, to get warrants to arrest Wylie, Chisum and J.M. Highsaw for the murder of Smith. April 21, Chisum sent a Mr. and Mrs. Gray, who lived at Seven Rivers, to Mr. Beckwith's house to tell the women and children to leave, that he did not want to kill them but that he wanted to kill the thieves that were in there. Mrs. Stafford, Miss Helen Beckwith and two of Mr. Beckwith's younger children were all the family who were present at Mr. Beckwith's. They refused to leave the house as they would not trust themselves to Chisum's men. Mrs.

Stafford replying that she had all the protection she wanted at home. April 22, Chisum advanced to carry his threats into execution telling his men they would try to take the walls on the other party, and kill them all who were in there. They continued shooting in about 700 or 800 yards distance, the besieged returning the fire. Chisum's men lost courage and would not advance any closer saying they were not going to get killed for $30 a month, that they had hired to herd cattle and not to fight. Chisum then raised a flag of truce and sent in a man to tell Mr. Johnson to come out, that he wanted to compromise and let everything go on as it was. Johnson told him he would not compromise for any person but himself, as there were men there whom Chisum and Wylie were owing money to and they wanted their money before the business was settled. Chisum then said that he would turn the water on and give them back their horses. He withdrew his men and went down to the river to round up his cattle, along with Wylie. They were going to round up together and separate their cattle when they got above Wylie's ranch. There John Chisum took the smallpox. May 7 Buck Powell having got back from Mesillia with the warrants, I, with a posse of fourteen men, rode down to Chisum's and Wylie's camp to arrest Chisum, Wylie and Highsaw. They, having learned that we were coming after them, had gone to South Spring River Ranch which is fortified against Indians, there to resist the law and fight if necessary. We started there in pursuit, after seeing John S. Chisum who was still too sick with the smallpox and could not be moved. On May 10, got to Chisum's ranch. Wylie, seeing we were there, sent out a man by the name of Charles Moore, whom we had seen at the cow camp, with a note stating there were men in there who did not want to fight and if we attempted to take the place some innocent persons might be killed. I sent him an answer that we did not want to kill any persons, that I had warrants to serve and I was going to serve them, that there was no use resisting. Wylie sent word back that he would meet Buck Powell and R.W. Beckwith halfway to the house and where we were camped, and he would pay them what he was owing them. They went and met him, and he told them to make out their bills and he would give them checks on the bank for the amount. He then told me that Highsaw was willing to stand his trial but he would not give himself up to my party as he was afraid they would kill him, if I would come up

> myself or with one man he would become my prisoner. When I got there he was gone, having gone the back way.
>
> Wylie then told the men he was owing to come down and meet Chisum who was coming up to his ranch, and he would get checks for their money. We all went down and saw Chisum, and Wylie got the checks and paid the men, and I then served the warrant on John S. Chisum and placed him under bond as he was still sick and every man went home about their business.
>
> Andrew Boyle, Deputy Sheriff.

Fulton wrote that Boyle's report was dubious and "excited little contemporary protest."[10]

Chisum ignored Boyle's account of the affair and some several months later the *Mesilla Independent* carried a letter from Mormon settlers living in the Roswell area befriending Chisum who had been their good neighbor for the past ten years. They wrote that Boyle's letter "in spirit and in fact, was devoid of truth as the lie which had been nailed." Part of the letter follows:

> The tone of the letter and the impression which it is intended to convey, is that Chisum wishes to monopolize two or three hundred miles of the Pecos River for his stock range, opposing immigration, persecuting newcomers, and using every endeavor to drive them from the country. This slander, we find, is being industriously circulated by Chisum's enemies, most of whom are indebted to him for their substance, and are using these subtle means to prejudice public sentiment against him for no other reason than that he objects to their longer preying upon him, killing, driving off and selling his stock.

This letter represented twenty-four persons living in the vicinity of Chisum's ranch. It was dated, Rio Pecos, Lincoln County, New Mexico, September 6, 1877.[11]

Chisum gradually recovered his strength after the smallpox. He was one of the victims of an epidemic that struck New Mexico that year. The disease spread over the country and was so serious that court was adjourned for several months in Las Lunas and Socorro. Eighty persons died from the disease in Las Vegas, New Mexico, in 1877. Chisum carried scars on his face the rest of his life.[12]

Chisum men were not the only ones who lost their lives during this period of untamed New Mexico history. In her book, *My Girlhood Among Outlaws*, Lily Klasner wrote that John B. Slaughter, who later served as sheriff of Cochise County, Arizona, and operated a large ranch on the border of New Mexico and Arizona camped one time near Chisum's ranch on his way back to Texas. He was told at Chisum's store that two men, Gallagher and Boyd, were also camped in the neighborhood and were asking about him. Slaughter remembered them. They had stolen cattle from him in Texas. Several days later they rode up to Chisum's store, guns cocked, and Slaughter recognized them. "Hold up," he called out, "don't come any nearer or I'll kill you." The men, half drunk and seemingly looking for trouble, advanced on their horses, guns cocked. Slaughter, who had leveled his gun on a box in a wagon to steady his shot, fired. He hit Gallagher in the thigh. Boyd spurred his mount and escaped. Gallagher later died from the loss of blood but Slaughter talked to him a few minutes before he passed out, telling him that he would never let a man ride up on him armed. Gallagher told Slaughter that he had made a mistake, that he had been drinking and had come to the store to talk things over with Slaughter. He took a ring off his finger and asked Slaughter to send it to his wife in Texas. Slaughter was not questioned about the shooting since the killing was done in self-defense.[13]

Chapter 8

Causes of the Lincoln County War

Many writers have pondered the Lincoln County War, offering their individual versions of what actually caused one of the bloodiest vendettas in southwestern history. In the views of ten different writers, detailed as follows, a common thread of agreement is apparent — that the feud grew out of the Chisum-Murphy clash.

> Among the strong characters in Lincoln county was John Chisum — "Lord of all he surveyed except for the political power and business influence of Major L.G. Murphy — a power to be reckoned with in the crude politics of the day." His cowboys were as accomplished with firearms as any of Chisum's Texans. When a sheriff's posse composed of Murphy's henchmen murdered Tunstall, the Lincoln County War began.[1]

> The Lincoln County War was caused solely by a combination of circumstances arising from the efforts of Mr. Chisum to detect parties whom he suspected of carrying on a wholesale system of robbery of his herds. There was a lawless element in Lincoln County at the time, and when a dispute arose between promi-

> nent men this class was soon found arrayed on one or the other side. This was the case in this instance and little troubles arose between these partisans which led to bloodshed and the friends of each of the parties took sides. More killings resulted and soon nearly every citizen of the country seemed to be interested in the matter — not from choice but from necessity.[2]

> The war grew out of alleged thefts from Chisum and others by employees of Major L.G. Murphy and broke into savage conflict February 18, 1878, with the killing of J.H. Tunstall, a friend of Chisum and involved most of the cattleman's partisans.[3]

Probably Robert N. Mullin, who edited Fulton's book *History of The Lincoln County War*, expressed it best in his introduction:

> The Lincoln County War was essentially a struggle for economic power in a land where hard cash was scarce. Federal government contracts for the supply of provisions, principally beef, for the military posts and for the Indian reservations, were the grand prize. The competition for these contracts was bitter and frequently ruthless. Since the early 1870s Lawrence Gustave Murphy had been the Lincoln County subcontractor for William Rosenthal and the political clique at Santa Fe which enjoyed a near monopoly in supplying the government with beef, even though neither Rosenthal nor Murphy himself then raised or owned any significant number of cattle. They were challenged by John H. Chisum, owner of the largest herds of cattle in the territory, who declined to do business with Rosenthal but instead bid direct on the beef contracts. Thus began Chisum's struggle with Murphy and his successors along with their backers at Santa Fe — a struggle out of which grew the Lincoln County War.[4]

> The original cause was the rivalry existing between Lawrence G. Murphy and John S. Chisum at that time the leading cattleman at Lincoln and in the Pecos Valley respectively. Both were furnishing cattle for the Indian agency and each accused the other of stealing from their respective flocks. The rivalry of these two men was the basis of the war, though the acts and depredations in which the sympathizers of these two principals were involved may have brought on the actual crisis.[5]

Governor Lew Wallace's account of the war does not mention Chisum by name, but there is no doubt about whom Wallace referred, as can be seen:

> A rich stockraiser from Texas came up to New Mexico and in a short time had $200,000 worth of cattle. He settled on the Pecos River and in a little while had succeeded in driving away almost all the small grazers. To retaliate, they began stealing from him. Now, they do not steal in that country as they do farther east, but they drive off large herds of cattle, sometimes 500 at a time. To protect himself, the Texan went down into his native state and recruited about seventy men, murderers, thieves, and dangerous men of all classes, together with a number of sharpshooters and buffalo-hunters. His enemies seeing these war-like proceedings, banded together in common defense and the result was open war.[6]

> It was inevitable that Chisum, the cattle king of the Pecos Valley and Major L.G. Murphy, "Lord of the Mountains," would clash. Competitors in business, their kingdoms joined and jealousy and hatred occurred. The climax would be the Lincoln County War, "bloodiest vendetta in the history of the Southwest." [7]

> The feud was started by cattle stealing, murders of numerous men, and the indifference or corruption of the governor of the territory. As the feud worsened, two factions opposed one another. The faction we knew to be lawless and which was later so proven, was led by Mr. L.G. Murphy, a rancher who stole cattle from his neighbors, who also owned some stores in Lincoln and bitterly resented the opening of stores by others. Mr. Murphy, with his associates and guns, was powerful andcontrolled the corrupt government officials. The other group was led by John Chisum, A.A. McSween and J.H. Tunstall, all ranchers with businesses in Lincoln, and who, with others, had often been victims of the Murphy group.[8]

In the foreword of his book *Violence in Lincoln County*, Keleher wrote "Ghostly embers of that conflagration still exist in the recollection of Lincoln County pioneers," and expressed his opinion of what caused the war:

> . . . Men in high places in civil and military life schemed, connived and conspired to accomplish unworthy objectives, selfish men, ambitious for wealth and power, turned arrogant and hostile in their endeavors; ignorant men, acting in utter disregard of the rights of others, blindly followed stupid leaders along trails of dishonor which led to disaster.[9]

The Lincoln County War was a contest between Chisum and Murphy with the Pecos Valley as the pawn.[10]

Six-shooters ruled the day for the next few years. Tempers flared and bullets flew. People assumed several hundred persons were killed in the feud, but as Emerson Hough wrote in *The Story of the Cowboy*, "No one will ever know until the tally-book of the recording angel shows how many skeletons are lying out in the mountains of the calm southwestern country." Hough wrote that this "bloody, partisan contest" took place before modern communications reached the remote Territory of New Mexico, but if the Lincoln County war could have been reported, "it would have pictured the bravery, determination, faithfulness and hardihood of many, and the fiendish cruelty, lust of blood and the insatiable vindictiveness of others." [11]

Did John Chisum play the role in the Lincoln County War which some of his contemporaries as well as several writers and historians have ascribed to him? This is another question about Chisum which may never be answered. It was known that cowboys who had formerly worked for him did take sides, lining up with opposing factions. Chisum was an intimate friend of McSween and Tunstall, his name was associated with the war. In those days a man's word was a point of honor and Chisum could have made verbal agreements that were carried out. However, few documents pertaining to the war bore his signature. *The Grant County Herald*, June 8, 1877, reported that Chisum was paying four dollars a day for a man and his rifle and intended to drive everybody else out of the country.

George Anderson, who wrote a history of New Mexico, said Chisum was not a participant in the conflict and "at no time from the killing of the first man to the end of the war was any man employed by him engaged in any manner in the outrages referred to with his sanction or permission." Anderson also wrote that Chisum did not take any part in the bloody scenes instigated by Billy the Kid in his revenge for the killing of Tunstall.[12]

Maurice Garland Fulton wrote in his book *History of the Lincoln County War* that "although Chisum himself kept his distance from the actual scenes of battle, his landholding strength, his shrewdness and his discreet but powerful support of the McSween faction, were significant factors in the Lincoln County war.[13]

After the buffaloes had all been killed in the West Texas area, Frank Collinson, the English cowboy and buffalo hunter, set out for Lincoln County. He had heard that John Chisum was paying four dollars a day for a man and his rifle. (Collinson must have read the *Grant County Herald's* story.)[14]

When Lincoln was first settled by Spanish pioneers in the 1850s it was called La Placita, "the little town by the pretty river," where life was pleasant and hospitable. The name was changed to Lincoln in 1869 and the village became the county seat of Lincoln County, the largest county in the Territory. Little did its residents realize that eleven years later their village would become the scene of a bloody war unequaled in the Southwest, a war leaving a scar which even today has not entirely healed. Many events led up to and climaxed this frontier feud.[15]

John Chisum, a newcomer and "outlander" quickly established himself as the largest rancher in the territory. He was envied and hated by L.G. Murphy and his gang who until Chisum's arrival had been the unchallenged leader of Lincoln County.[16] Murphy was a power broker who drew his influence from myriad sources. He was, for example, a friend of Thomas B. Catron, New Mexico's then attorney general. He had other political allies in high places, including Governor Samuel B. Axtel. Juan B. Patron, a leading Spanish-American citizen swore that he had heard Murphy say on one occasion when demonstrating his "rule or ruin" policy to break up a political convention, "you might as well try to stop the waves of the ocean with a fork as to try and oppose me." (Affidavit by Patron, July 1, 1878, Keleher, *Violence in Lincoln County*, p. 249). In short, Murphy had the support of the "Santa Fe Ring," a group of fat cat politicians headed by Catron who wielded the power in political activities throughout the Territory. There was whispering about the power of the ring, its manner of first threatening and frightening people, then acting, perhaps with violence and always with severe results.[17] If the Territory itself had a large ring, Lincoln County had a small ring run by L.G. Murphy and his associates. Murphy, himself, was president of the First National Bank of Santa Fe, as well as a rancher and land speculator.

Fulton wrote that the existence of the secret political society was hard to prove but Chisum believed it existed. In 1878, when men he believed were members of the ring tried to ruin him by pressing collection of packinghouse notes, Chisum wrote his opinion of the ring:

> I know it is said by many of the citizens that there is a ring in Santa Fe, but the governor said in his last message that there was no ring. I am of the opinion, if there was a ring, there would be some show to get through it or to get on the outside or the inside of it. But the thing they have got up there is perfectly hard and solid, it cannot be penetrated from either side or end, and it is perfectly solid. When this substance strikes a citizen it goes right through him and leaves him in such a condition that he never recovers from it. It is so powerful that it not only ruins citizens but sometimes it ruins whole counties.[18]

Among the new citizens of Lincoln about this time was a distinguished young lawyer, Alexander A. McSween and his beautiful and cultured wife, Susan. They were married in Kansas in 1873 and came west in 1875. McSween was well educated and as he grew older was religiously inclined. He had thought about being a minister, but ambition for fame and fortune persuaded him to study law. His parents were Scottish. He was born on Prince Edward Island.[19]

Arriving in Lincoln at the same time, was a wealthy young Englishman, John H. Tunstall, who had money to invest. He was the only son of John Partridge Tunstall of J.P. Tunstall and Company, a mercantile firm in London. John immigrated to California in 1875, planning to invest in a sheep ranch, but hearing that New Mexico had a better range, he came to Lincoln, filed for land on the Penosco and later on the Rio Feliz.[20]

It seemed natural that McSween and Tunstall should become fast friends. Both were well educated, ambitious and young. They soon formed a partnership and, in competition with Murphy, opened a mercantile store and a bank. Chisum was interested in the bank and his name was listed on the bank's letterhead as president. He had become a staunch friend of the two younger men. His friendship with them aligned Murphy, Dolan and Riley and many smaller ranchmen against Chisum.

Major Emil Fritz was a partner in the Murphy store before his death in 1874, and Murphy later took J.J. Dolan and John H. Riley as partners. In 1877 Murphy resigned from the firm and the two partners, Dolan and Riley, continued the business under the name of Jas. J. Dolan and Co.[21]

Murphy was furious when he saw his power and wealth being threatened by his competitors. The McSween-Tunstall general store was drawing a major part of the local trade. Murphy swore vengeance against McSween. The latter had worked for Murphy when he first came to Lincoln, but resigned when he realized many of Murphy's activities displayed a notable lack of integrity. When McSween was immediately hired by Chisum, Murphy's anger intensified. Chisum, McSween and Tunstall formed a closely-knit association and prospered in their banking and retail business.

Meantime, Murphy nursed his lust for revenge, scheming in every way to get the best of the three newcomers who had dared challenge the established order. As it turned out, Murphy succeeded very well.

A ten thousand dollar insurance policy also sparked the

flames that helped to start the Lincoln County war. Murphy's first partner, Emil Fritz, died and left such a policy to be paid to his sister and brother. Murphy held both the will and policy saying the policy was security for debts owed him by Fritz. The Fritz heirs asked McSween to intervene in their behalf. They instructed him to go east to collect the policy for them. He was able to get possession of the policy but not the will.[22] He then went to New York, paying his own way, and was able to successfully settle the claim in favor of the Fritz heirs. On August 1, 1877, Donnell Lawson and Company, the New York insurance firm, notified McSween that the company was crediting him with $7,148.94, the amount remaining after deducting expenses accrued by McSween, the Fritz family and the insurance company.[23]

When Murphy and his partners heard the news that the insurance claim had been settled, they were determined to get their hands on the money one way or another. If they could not get the actual cash they would take its equivalent in merchandise from the McSween–Tunstall store, or in horses from the McSween–Tunstall partnership ranch.[24]

Chapter 9

The Killing of Tunstall and Sheriff Brady

One Sunday Chisum rode into Lincoln to visit the McSweens. At the same time two ranchmen from the Ruidoso rode into town. They forced the saloon keeper to open the bar and soon were roaring drunk. Thus fortified they got back on their horses and rode to the McSween home, firing wildly at the house and shouting that Chisum's corpse be turned over to them or they would burn the house down. A window was shattered and furniture riddled by bullets. Before more serious damage was done Sheriff Brady came to the rescue.[1] This was an example of the hatred and envy that some of the smaller cowmen felt for the Cattle King of the Pecos.

Stories of violence were reported daily as hate and vengeance spread over Lincoln County. Bad men gravitated to the area, contributing to the tension and paranoia. Jesse Evans's gang was present and riding over Lincoln County in 1876 and 1877. One of the members of Evans's gang, was Jim McDaniels, a former Chisum cowboy who had quit in a dispute over indebtedness. He often bragged about getting even with "Old John." [2]

Tunstall had tried to employ honest and competent men to work on his newly-founded ranch on the Penasco. Dick Brewer had been hired as foreman and was in every way satisfactory and dependable. The others were chosen from "the flotsam and jetsam in Lincoln County," and among them were William Bonney, later known as Billy the Kid, John Middleton, Fred Waite and several others.[3]

Against this backdrop, Murphy continued to plot against Chisum. Convinced that McSween would not turn over the insurance money in lieu of the debts he claimed Fritz owed the firm, Murphy's next move was to order Sheriff William Brady, known to be "the Murphy sheriff," to levy on the Tunstall ranch. It was rumored McSween had put his property in Tunstall's name to keep his creditors from getting any of it. Tunstall and McSween were known to be partners in the ranch.[4]

In 1878 Sheriff Brady organized a posse headed by Deputy Sheriff William S. Morton to ride to Tunstall's ranch. The posse included such notorious outlaws as Tom Hill, who had previously threatened Tunstall's life, Frank Baker and Jesse Evans. Before setting out, they apparently put iron into their resolve with a few stiff drinks.[5]

According to Frank W. Angel, a New York lawyer who was sent to New Mexico to investigate Tunstall's death and other violence in Lincoln County, when the posse reached the ranch they found Tunstall gone. The posse set out in hot pursuit, overtaking Tunstall and his group of cowboys, about thirty miles from the ranch. They were driving a herd of horses with them. Tunstall was fired upon. He left the horses and tried to escape, but was killed by the mob. "Who shot him will never be known," Angel wrote, "but there is no doubt that Deputy Morton, Jesse Evans and Hill were present and saw the shooting and that two of these persons murdered him, for Tunstall was shot in two places, in the head and breast." [6] Tunstall's body and that of his fine horse, also killed, were left in the road.

"There was no object for following after Tunstall except to murder him," Angel wrote, "because they had the horses which they desired to attach before they commenced to pursue him." Angel believed that Tunstall was murdered in cold blood and not shot in attempting to resist an officer of the law, as members of the posse reported. On an earlier occasion, Evans and Hill were arrested at the instigation of Tunstall and they had hated him ever since. Angel wrote that such enmity meant death. Angel concluded his report by writing that Morton and Hill were killed later. The only survivor was Jesse Evans, "notorious outlaw, murderer and horse thief." [7]

Angel spent several months in New Mexico thoroughly investigating Tunstall's short life in the Territory. He described the young Englishman as straightforward and honest in his business dealing, adding that he had "almost overthrown a certain faction of said country that was plundering the people." He credited Tunstall as having been instrumental in the arrest of certain notorious horse thieves and in exposing embezzlement by some Territorial office holders. Tunstall incurred the anger of persons who occupied positions of power in the country and abused that power for private gain. The enmity of those persons resulted in his death.[8]

Violence would soon flow like an epidemic of smallpox after Tunstall's death, but first his funeral was conducted with most of the countryside attending. Services were held in the open with Dr. Ealy in charge. His remarks were translated into Spanish for the benefit of Tunstall's Spanish friends. A small organ, belonging to the absent Mrs. McSween, was carried out to the ground behind the Tunstall store. Men and women stood around it, "deeply aroused over the brutality and lawlessness which had brought on the murder of the kindly and ineffectual young Englishman." [9]

Billy the Kid and Dick Brewer, Tunstall's foreman, had been riding with Tunstall on the day of his death but had ridden off to chase a wild turkey. Hearing the gunshots, the two returned to the road and found the body of Tunstall and his horse. A distant dust cloud floating in the air and the echo of voices, now drunken and hilarious, were all that were left of the posse. Billy was deeply affected by Tunstall's murder and swore vengeance on every man in the posse. Tunstall was his idol of manhood and now he was dead.[10] "The senseless brutality of the cold-blooded murder of Tunstall was to exert a profound effect on Billy the Kid throughout the remaining days of his tragic young life. Nothing had ever touched him so deeply save only the death of his mother. It was largely to shape his destiny and decide his fate." [11]

Billy once told Frank Collinson that Tunstall was the only man who had ever treated him like a human being.[12]

Billy joined a posse headed by Brewer to run down and arrest the murderers of Tunstall. Deputy Morton and Frank Baker were later caught after a running battle and forcibly taken to the Chisum ranch for the night. On the following morning, March 9, 1878, the party headed for Lincoln. Trouble arose among the group en route and three of the posse, Deputy Morton, Frank Baker and a man named McClesky were killed. McClesky, who had been riding with Morton and Baker, was shot as he tried to defend them.[13]

There was still one more member of the Tunstall killers that Billy the Kid swore to get. He was "Buckshot" Roberts, ex-soldier and ex-ranger, who lived near Blazer's Saw Mill above Tularosa. He was known as "Buckshot" since he carried a load of lead in his right shoulder and could not lift his rifle above hip level. When the gang rode up to his cabin, including the Kid, Brewer, George and Frank Coe, Charlie Bowdre, Dock Middleton, Scroggins, and Steve Stevens and several others whom "Buckshot" did not know, he bravely faced them all. When they began shooting around the corner of his cabin, "Buckshot" opened fire with his rifle, shooting from his hip. He hit Middleton, shot a finger off George Coe's hand and jammed his rifle muzzle into Billy's midriff, but it did not fire. Charlie Bowdre then shot "Buckshot" through the body and he staggered into the house. Snatching a buffalo gun he shot through the window striking Dick Brewer in the head and killing him instantly. The Kid and his gang swiftly rode away.[14] "Buckshot" died from Bowdre's wound.

Intimidating letters were sent to the Territory newspapers during those days. They fine-honed the tension in the air. Chisum was blamed, some said falsely, for much of the trouble. *The Grant County Herald*, June 8, 1878, published the following:

> I know that Chisum would be glad to see me killed and so I just left my crops and sailed out. Lots of men are leaving the country in the same way, because they say that McSween and . . . Chisum are paying four dollars a day for a man and his rifle and intend to drive everybody out.

Rustlers continued to steal jinglebob cattle and although Chisum's cowboys were loyal and courageous there were not enough of them to stand guard over Chisum's vast valley kingdom. Chisum busied himself collecting evidence which enabled him to swear out warrants for men he suspected of stealing his cattle. Sheriff Brady arrested some of the suspects, but they soon escaped the Lincoln Jail, or were freed through their own efforts or that of friends who rode into town making a noisy show of six-shooters and Winchesters. Sheriff Brady held warrants for Billy the Kid and his companion for killing Morton, Baker and Roberts. The sheriff left the Murphy–Dolan store one morning with George Hindman and J. B. Mathews en route to the courthouse to announce that there would be no regular April term of court. In the rear of the Tunstall–McSween store that stood midway between the Murphy–Dolan store and the courthouse, there was an adobe corral whose east side extended behind the store building, commanding a view of the street. Billy and his warriors had cut grooves along the top of this wall to rest their rifles. As Brady and his companions came into sight they were met with a volley from the adobe fence. Brady and Hindman were mortally wounded, but Mathews found shelter behind a house on the south side of the street. Billy leaped over the adobe wall and ran to the fallen officer and picked up the sheriff's gun. As he did so he was fired upon by Mathews and the ball shattered the stock of Brady's rifle and slightly injured Billy in the side. He ran for shelter to the McSween home. Many of the Lincoln citizens turned against Billy after the cold-blooded murder of Sheriff Brady and he and his warriors were looked upon as outlaws in every sense of the word.[15]

Tunstall's death was soon forgotten in the excitement caused by the murder of Sheriff Brady and his deputy. Feelings ran high. McSween fearing for his life, hid for a while in the Bottomless Lakes area and Mrs. McSween stayed in the Chisum ranch home.[16]

At times McSween and his friends, including William H. Bonney, who would become better known as Billy the Kid, rode into the Chisum corrals to spend the night on the ranch. Sallie Chisum, a niece who resided with her Uncle John at the time, wrote an account of such a visit in March 1878. Morton and Baker were being held prisoners at the ranch. Chisum was not at home when McSween and his companions arrived. Sallie wrote that all of the men were fed and provided with sleeping quarters. Sallie's room was turned over to the prisoners since it was the only one in the house with no outside exposure. Sallie reported that a guard, armed to the teeth, watched the prisoners all night to prevent any possibility of escape.[17]

William H. Bonney, a buck-toothed, beardless youth, was not so different in appearance from the youth of the 1970s counter culture, with long blond hair. He fell for pretty Sallie. After supper Billy wanted to go fishing and Sallie's brother, Will, hunted up hooks and lines and the two boys fished from the earthen dam a hundred yards from the house. Sallie was destined to see Billy from time to time and she said he was always a perfect gentleman around the ladies. That first trip to Chisum's ranch marked Billy's arrival in jinglebob country, and he later made the acquaintance of the King of the Jinglebobs.

Deputy Morton and Baker feared they might be killed before arriving back in Lincoln and Morton wrote a letter to H.H. Marshall, a Richmond, Virginia lawyer, which he mailed in Roswell, five miles northwest of Chisum's ranch the following morning. That letter was later published in the *Independent*, April 18, 1878. Among other things Morton wrote:

> . . . I have heard that we are not to be taken alive to Lincoln. If anything should happen, I refer you to T.B. Catron, U.S. Attorney, Santa Fe, N.M. and Colonel Rynerson, District Attorney, La Mesilla, N.M. They know all about the affairs . . .

There was some truth to this *Mesilla News* item of July 14, 1878:

> John S. Chisum, it is reported, furnishes . . . McSween's boys with fresh horses to ride after they run awhile from the sheriff—Chisum's house is also opened to them to use as a fort against the officers of the law . . .

George W. Peppin, who had been appointed sheriff of Lincoln County after Brady's death, also reported that summer that McSween's forces had been "fed, clothed, and horses exchanged for fresh ones, by the Chisums." [18]

J.K. Millwee, one of Chisum's faithful cowboys, told about an all-night meeting in the jinglebob ranch house in 1878 during the feuding days. This preceded Tunstall's death. Gathered in the large living room of Chisum's South Spring Ranch home, were Millwee, who had been with Chisum six years, Frank McNabb and Dick Brewer, who were growing old in Chisum's service, Jim Sanders, bronc rider, Harvey Morris and Francisco Semora, line riders. Tunstall, the newly arrived Englishman, was staying with Chisum until he could hire serviceable men to help run his recently-established ranch on the Feliz River. Then, of course, there were Alexander McSween, Lincoln's only lawyer, and that young man, merely a boy, named William Bonney. Bonney drifted into the Pecos country after having been in Old Mexico. Little was known of him, but it was whispered he had killed a man when only fourteen years of age and had been avoiding the law. On this evening he sat in a far corner of the room, his long blond hair falling to his shoulders, his blue eyes studying the faces of each man present, while his long, slim fingers that would later become so adept at pulling triggers, plucked the strings of a banjo.

Chisum had the floor. He knew that some of his men would choose sides in the conflict and told them they were free to leave his employ if they wished. If not, they could remain on his payroll. Millwee said that he thought about the proposition all night. He knew that if a man chose a side he would have to carry a gun. He did not believe in carrying a gun. He was a man of peace. He decided to leave and said to Chisum, "I care nothing about your feuds and gun fights. I have learned that one killing merely calls for another." Baker and Morton, then working for Chisum, decided to leave and later entered the services of Murphy and Dolan, Chisum's enemies.

Milwee, who rode back to Texas, managed several ranches and later owned his own spread, never forgot that memorable night in the Chisum ranch home. He said that all of the men gathered there, with the exception of Chisum, died violently. Tunstall was killed by a drunken mob. Frank McNabb and Jim Sanders were shot by a sheriff's posse in Benito Canyon near Lincoln; Billy Morton and Frank Baker were killed by Billy the Kid while trying to escape capture by Tunstall's supporters. Harvey Morris and Francisco Semora were killed while trying to escape from McSween's home. McSween was killed as he walked unarmed out his front door. Billy the Kid was killed by Pat Garrett at Fort Sumner in 1881.[19]

On one occasion McSween and his followers found refuge in Chisum's ranch home behind fort-like four foot thick adobe walls.[20] They were followed by deputy sheriff Marion Turner and his posse. The posse halted near the jinglebob corral but did not attack the house. James Chisum, Sallie's father, and Billy the Kid were on the flat roof of the house keeping an eye on the posse. No shots were fired and being unable to get to the barns or house, the lawmen left. Some of them returned a few days later, but James Chisum met them at the corral to tell them there were women and children in the house and warned the men about firing their guns. The group rode away without further ado.

Territorial newspapers began to carry incriminating articles that stirred up old quarrels. Murphy and Dolan sent several such epistles to the *Mesilla News* and *Santa Fe New Mexican*. Such articles were even published as far away as Arizona. The following story appeared in the *Arizona Miner*, Prescott, June 7, 1878:

> Lincoln County, New Mexico, seems to be undergoing a reign of terror. A lawyer by the name of McSween is largely responsible, according to reports from Murphy and Dolan. They also accuse J.S. Chisum, the cattle king, of being in with McSween and other parties making to defraud Murphy and Dolan and other good men. Dolan, in a letter to the *New Mexican* states Chisum wants to control the contracting business of New Mexico and Arizona.

Chapter 10

The Arrests of Chisum and McSween

Meantime, Chisum's troubles multiplied. Smaller ranchers accused him of driving off their stock with his own herds and Chisum in turn accused them of driving off his stock with their outfits. Charges and countercharges flew. Murder continued to occur frequently, and hatred, like an open sore, festered and flared between opposing factions.

Lawrence G. Murphy had become Chisum's great nemesis in the Territory. His "cattle ranch," was first said to be on the Carrizoza plains, but Murphy later moved it to the Seven Rivers country, "close, uncomfortably close to Chisum's jinglebob holdings." [1]

It was generally believed that Murphy and Company fulfilled the contract they had with William Rosenthal by buying stolen cattle which were kept at a cow camp near Seven Rivers. There they did business with cattle rustlers. Jesse Evans and his gang of outlaws, rumored to be the thieves, were reportedly on a secret payroll of Murphy and Company.[2]

Chisum also was haunted by an old fear centered on

some notes he had signed forming a brief association with Wilber, Chisum and Clark, an ill-advised packinghouse venture in Arkansas. The notes dated to 1867 when Chisum was ranching at Bolivar, Texas. He believed the notes unjust and was determined not to pay them.

There is little doubt Chisum had those notes in mind when he made the deal to sell out to Hunter, Evans, and Company, and again when he advised Wylie to keep his herd in his own name even though Chisum had bought the cattle. No doubt the smaller registered herd which Chisum kept for himself was in the names of his brothers, Pittser and James, though Chisum himself remained boss for the ranch.

After the sale of his cattle to Hunter, Evans and Company was announced in the public press along with the huge sum of money he was reported to have received from them, Chisum began to be tormented by the prominent Santa Fe law firm of Catron and Thornton. Catron, himself the most outstanding attorney in the Territory and known by Chisum to be a close associate of James Murphy, was said by historian Fulton to have "a dominating personality, not to say ruthless, and during the 1870s as well as a decade and a half longer, he was virtually dictator of New Mexico, politically, commercially and professionally." [3]

Catron was no stranger to Chisum, having previously confronted him over petty debt matters in court, according to Lincoln County court records. In 1875, Chisum was sued for $2,614 in taxes claimed to be due the Territory. Chisum was represented by McSween, while Catron prosecuted the case. Catron filed several minor suits against Chisum, "mostly for small amounts said to be due on license and tobacco taxes." [4]

One writer wrote of Catron, "It was generally believed that his personal aim was to cripple Chisum's prestige and therefore endanger his range claim so as to occupy parts of it to his advantage." [5]

Events were catching up with Chisum and, in an ancillary way, with his friend, McSween. It would soon be clear, also, that the fine hand of their common Lincoln County enemy, James Murphy, would be seen tying collaborating knots in a snare Catron apparently hoped would bring down Chisum once and for all.

Matters climaxed in December, 1877, when Chisum and McSween, accompanied by McSween's wife, set out by the commonly used "ambulance" carrier for Trinidad, Colorado, where they planned to entrain for St. Louis. The timing of the trip was purely coincidental. The men had business to attend to in St. Louis and Mrs. McSween was simply going along for the trip. They reached Las Vegas, New Mexico, on Christmas Day, where, to their surprise, they were taken into custody by the sheriff. The lawman had been instructed by attorney Catron to place the travelers under a forty-eight hour delay until warrants could be issued to prevent their leaving the Territory.

Chisum was held on a charge of resisting arrest.[6] To assure that McSween would be held in the Las Vegas jail, Murphy had sent his partner, J.J. Dolan, to Las Cruces to talk Mrs. Scholand, the sister of Fritz, into signing an affidavit accusing McSween of leaving New Mexico for good with the insurance money. Dolan persuaded Mrs. Scholand, who neither spoke nor wrote English well, that McSween planned to use for himself the insurance money that had belonged to her late brother. Dolan delivered the affidavit to District Attorney W.L. Rynerson, which resulted in McSween's arrest.[7]

A few days later, McSween was returned to Lincoln and questioned by Judge Bristol about non-payment of the insurance claim he had collected for the Fritz heirs. McSween would be tried in April for embezzlement, but his case seemed a thinly veiled frame-up, especially in view of his insistence on returning to face the charges.[8]

While in jail, Chisum wrote the following account of his and McSween's arrests:

> In December 1877, I started for St. Louis, Mo. on some business and got to Las Vegas on the night of Dec. 24. Upon arrival there I found T.B. Catron had telegraphed to Las Vegas to know if Chisum and McSween had passed Las Vegas yet. The sheriff

> commanded us not to leave the city. I was traveling in company with McSween and his wife, all going to St. Louis on business and of course in a hurry as all travelers are. So the McSweens and I waited for forty-eight hours and no warrant came for the arrest of either one of us. The sheriff held us as long as he could without papers. So we hitched up our horses to the ambulance and started for St. Louis. After going half a mile we were overtaken by the sheriff and stopped on the public highway. The sheriff had thirty or forty men with him, some armed with pistols, some with rocks, some with clubs, but all making a desperate charge upon us and surrounding the ambulance we were traveling in. I was jerked out head foremost and fell upon my face on the hard road. I was then seized by the throat by one of the sheriff's men until I said to the sheriff, "Will you please be so kind as to loosen his hold." So I breathed once more of the fresh air of New Mexico that they brag so much about. McSween was also jerked out of the ambulance and dragged by a lot of the gang. Mrs. McSween was left sitting all alone crying in the ambulance without a driver or even protection. The sheriff and his party were somewhat excited. I looked the gang over and noticed one young man who had on a clean shirt who had just arrived at the scene of the excitement. I asked him if he would be kind enough to drive Mrs. McSween to the hotel, and the young fellow consented to do so and drove her there.[9]

Now Catron had Chisum where he wanted him, in jail. Catron had several aces up his sleeve. He could try Chisum on a promissory note matter, a cattle deal several years back with Charlie Goodnight, and on another ghost in Chisum's past — the packinghouse notes. He took Chisum to court to collect on a promissory note the latter had endorsed to Alexander Grzelachowski of Puerto de Luna for $3,000 which Chisum had refused to pay.[10]

The suit went to trial in neighboring San Miguel District Court. Catron won the suit, but apparently lost it on a technicality. When Catron tried to collect the money he had to file an affidavit alleging that Chisum had transferred his ranches to Hunter, Evans and Company, "with intent to hinder, delay and defraud creditors." Due to the affidavit, Catron, asked the court to issue the writ of *no exeat* against Chisum. *No exeat* was an early common law of England which meant "Let him not go out of the Kingdom." Chisum and his lawyer fought the case vigorously and won when the judge ruled that Catron failed to prove sufficient facts to justify the issuance of the writ.[11]

The relentless Catron also brought Chisum to account for cattle Goodnight had refused to buy from him some years back on grounds that he, Goodnight, believed them to be stolen.[12]

The case went to court in Las Vegas and when Chisum made his entrance in the courtroom with a long zinc tube he caused a stir of curiosity among spectators. Chisum was often seen on the home range with the same tube tied securely behind his saddle. Now, in court, he laid the cylinder on a table and began extracting from it powers of attorney from Texas cowmen who had given him the authority to sell their herds along with his own. Catron realized he had lost and asked for dismissal of the charges. Then Catron brought Chisum to trial on the old packinghouse notes, invoking an old law which demanded imprisonment when a person "against whom a judgment has been procured, refused to list his assets." Catron and Thornton prohibited Chisum from leaving the jurisdiction of the court.

William Rosenthal and others claimed to hold court judgments against Chisum. They had secured possession of the old Wilber–Chisum–Clark notes and were determined to sue Chisum for debt. It made no difference to them that Chisum denied the partnership and said his name was on the notes without his knowledge or consent. As a result of the suit, Chisum served some eight weeks in jail. Hoping to counteract his enemies, he wrote his own account of the packinghouse entanglement.[13]

In his account, Chisum said he became a partner in a packinghouse firm in Arkansas while ranching in Bolivar, Denton County, Texas. The company was organized to operate under the name Wilber-Chisum and Clark.[14] Chisum's responsibility would be to buy cattle and deliver them to the firm. He agreed to enter the partnership contingent upon

Clark and Wilber raising $66,666.66, and investing in a packinghouse, salt, barrels and other equipment necessary to the operation of the business. After this was accomplished Chisum would put up $33,333.33 in beef, provided the other men would not pack over 5,000 barrels during the first season. Chisum and Wilber wrote out a rough outline of their agreement, both signing and keeping a copy.[15]

According to his written account, Chisum had second thoughts about signing the agreement. Wondering if he had acted too hastily, he wrote to a friend in Arkansas inquiring about Clark's background and was advised to stay clear of Clark, that "he had nothing." [16]

Meanwhile, when he heard nothing further from either Wilber or Clark, Chisum assumed the venture was ended and in 1863 moved part of the herd to the Concho.[17]

Chisum spent several years in West Texas and in 1867 went on to New Mexico. In July of that year he trailed a herd up the Pecos, arriving in August at Bosque Grande. After settling permanently in the Territory, he learned that the packing company was using his name in its continued operation.[18]

> Then and not until then did I learn that Wilber and Clark had been packing beef at Fort Smith in the fall of 1867 under the names of Wilber-Chisum and Clark. One of my brothers heard that such a firm was packing beef at Fort Smith 300 miles distant and knowing where I was and fearing something was wrong, he went to Fort Smith and found out that what he had heard was true. They were in the mercantile business and a large sign over their store as well as over their packinghouse proclaimed that their bottles of cocktails were labeled 'Wilber, Chisum and Clark,' which they had made in New York. One of my brothers exposed the whole thing that evening, and the next day they were closed up and the swindle exposed. [19]

Chisum made a special trip to Austin, Texas, to discuss his problem with Governor Elisha Pease. The governor assured him that the law would not hold him responsible since he had been living out of the state and knew nothing of the firm's operation at that time. Chisum was sued in Texas on a few of the notes, but the court ruled he was not responsible for the contracts.[20] He had been living four years in Texas after the notes matured before pushing his herds farther westward.

When the packinghouse suspended business in Fort Smith some $80,000–$90,000 was due in promissory notes. The owners employed attorneys who advised Chisum that since he was the only member of the alleged partnership who had not taken the bankruptcy law he was being sued for payment.[21]

The packinghouse notes were all of eight years old when William Rosenthal of Santa Fe, merchant and beef contractor, traced the owners and bought the paper for a few cents on the dollar. As owner of the notes, Rosenthal filed suit against Chisum for payment. Chisum refused to pay and, to make certain Rosenthal would not get any of his property, transferred most of it to his brothers, Pittser and James.[22]

By early March 1878, Chisum executed his bond and was released from custody.[23] His brother, James, journeyed to Las Vegas to bring John back to the ranch. Chisum's enemies had profited little by his incarceration. No Chisum assets had become public and no money had been paid on the packinghouse notes, the Grzelachowski promissory note or the Goodnight matter.

It had been impossible for claimants to attach any of Chisum's property since the Lincoln County sheriff reported that Chisum had no property to attach. Nevertheless, the claimants were convinced that Chisum's property, worth several hundred thousand dollars, was being held by a third party under a specious title.[24]

Was Chisum faced with still another shocking experience and bitter disappointment when he settled with Colonel Hunter for his cattle? Several reliable writers have recorded such a story. Though some historians place little credence in the following incident, it does provide a different version of the outcome of Chisum's sale of his cattle business to Hunter, Evans and Company. It also demonstrates how, once unleashed, adversity and rumors of adversity pursued Chisum like the "hounds of hell."

The story goes that Hunter, who was represented by Catron, was doubtless filled in by the latter on the cattle Goodnight had refused to accept from Chisum. Hunter made use of the information by riding into Texas, and for a few cents on the dollar, buying the old notes Chisum paid West Texas cowmen for the cattle he delivered, along with his own herds, to Goodnight.[25]

The *Dodge City Globe*, April 6, 1876, reported that Jesse Evans, the Texas outlaw, was leaving Dodge City for the Pecos with a gang of fifty cowboys to gather up a herd of 20,000 cattle purchased from Chisum by Hunter, Evans and Company. After Chisum turned over the cattle to the men, the animals were driven toward Texas, the end of the trail ultimately being Dodge City.[26]

Hunter was said to have met Chisum in Catron's office where the final payment for the cattle was to be made. Several witnesses were said to be present. Placing a small hand satchel on the table, Hunter said, "Here is your money, Chisum." When Chisum opened the satchel he received a staggering financial blow. Instead of gold or greenbacks, he found the notes he had given the West Texas cowmen and which Hunter subsequently acquired. Thus, instead of the huge sum of money he had reportedly received when he sold out to Hunter, Evans and Company, Chisum was paid for his cattle with his old notes at par value.[27]

In his book, *The Trampling Herd*,[28] Wellman wrote that after his initial explosion of rage, Chisum might have rushed out, rallied his cowboys and set out after the disappearing herd to hold it. He could also have settled the matter in court later. However, another decisive factor intervened. Chisum knew the cowboys riding with Jesse Evans were among Dodge City's most experienced gunmen and they carried "swinging in the holsters at their side, certificates of their ability to hold the cattle against any but the most overwhelming force." As a consequence, Chisum had to stand by helplessly, watching the 20,000 cattle as they were driven toward the Texas border forever beyond his reach.

The story of Hunter's legal "rustle" of Chisum's cows was widely told and enjoyed in the Southwest.[29]

Chisum, it was said, offered big inducements to Billy the Kid and his gang to follow the herd and bring back the cattle, but Billy refused, knowing the Hunter and Evans cowboys were armed to the teeth and that the cattle had already crossed over into Texas.[30]

The foregoing account of Chisum's final payment by Colonel Hunter makes one wonder, especially in light of the new information contained in Mrs. Hunter's journal. Would Colonel Hunter have permitted his younger brother, David, and David's womenfolk to go to the Chisum ranch on such a responsible mission and for such an extended stay had he not himself thought well of Chisum? According to Mrs. Hunter's journal, the visitors shared a good opinion of Chisum and until the feud increased in tempo toward the latter part of their visit, the Hunters seemingly enjoyed a pleasant sojourn on the ranch.

To have paid Chisum with his own notes would have been a calculated act on Colonel Hunter's part, indicative of an absence of regard for Chisum and indifference to the certain end it would bring to their association. No evidence of earlier bad feelings between Hunter, Evans and Company and Chisum has been unearthed in research by this author.

Sallie Chisum, "a pretty fair-haired girl, full of life and ready for any kind of sport," as the Hunter women described her,[31] was living on the Chisum ranch when the Hunters were there. The women became intimate friends and corresponded for years after their separation. Sallie was a great favorite of her Uncle John, and was called "The Queen of the Jinglebob," by the cowboys. Sallie knew her uncle's personality and habits and loved him dearly. She told Colonel Jack Potter, a former Chisum cowboy, that the relationship between her uncle and Hunter, Evans and Company, was never strained as was rumored at the time. Surely, Sallie would have known it had Colonel Hunter paid John Chisum for the cattle he bought from him with Chisum's old notes at par.[32]

Chapter 11

Spring Term of Court Keeps Grand Jury Busy

The grand jury spent many wearisome days in the Lincoln spring court of 1878 investigating the recent and current crimes in the county. They brought indictments against Jesse Evans, George Davis, Miguel Seguro and Frank Rivers as principals in connection with the Tunstall murder along with James J. Dolan and J.B. (Billy) Matthews, a younger member of the Dolan firm. Among the principals only Jesse Evans could be found. He was arrested and placed under a $5,000 bond. Dolan and Matthews were also arrested and placed under $2,000 bonds.[1]

Indictments were brought against William H. Bonney (Billy the Kid), John Middleton, Fred Waite and Henry Brown in connection with the killing of Sheriff Brady and his deputy, Hindman. Charlie Bowdre was indicted for killing Buckshot Roberts. Since none of these men could be found, it was impossible to make arrests. McSween was freed from the criminal charge of embezzlement but he still faced the civil suit which had provoked the attachment proceedings. Dolan and Riley were also indicted for cattle stealing.[2]

The *Cimarron News and Press* of May 2, 1878, gave much publicity to the grand jury's report and concluded editorially that the indictments for cattle stealing and receiving stolen cattle, "would break up the arrangements heretofore existing and entirely destroy the usefulness of Captain Jesse Evans and his band in that section."

Dolan published the following notice in the *Santa Fe New Mexican* after the newspaper account of the grand jury report:

> The condition of affairs now existing in the country is such as to make it unsafe for the undersigned to further continue business as they have heretofore done. They take this occasion to assure their friends and the public that the suspension will be only temporary, and that they will resume when peace and quiet shall take the place of lawlessness, and order be restored in the country. Asking for continued confidence and hoping for a renewal of business relations, we remain, very resp. Jas. Dolan and Co.[3]

Fulton wrote that behind Dolan's statement was the desire of T.B. Catron to protect himself in connection with thousands of dollars in First National Bank, Santa Fe, notes he had endorsed in behalf of the Dolan firm. In return for his endorsement, Dolan and Riley had mortgaged to Catron virtually all their property, including the forty acres surrounding their places of business and 1500 head of "Arrow" brand cattle ranging near Black River on the Pecos.[4]

Soon after the grand jury report, Murphy, who was in poor health, left Lincoln for Santa Fe where he died October 19, 1878. Dolan accompanied Murphy to Santa Fe, but later returned to Lincoln, determined to again control his business, or at least to wind up his affairs. Riley, who had gone to Las Cruces, also returned to Lincoln to help Dolan reestablish their business.[5] Although they had suffered reverses they were determined to carry on the feud with McSween. Atanacio Martinez, the constable, was said to have overheard W.L. Rynerson, district attorney, tell Dolan, "Don't give up. Stick to that McSween crowd. I will aid you all I can and will send you twenty men. Stick to the fight and give it to those rascals." [6]

Meantime, McSween offered a $5,000 reward for the apprehension and conviction of the murderers of Tunstall.[7]

The above-mentioned indictments must have been unusually good news for Chisum, who was back home from prison when his good friend, McSween, was tried for embezzlement. McSween was cleared. The jury's verdict was as follows: "We fully exonerate him on the charge and regret that a spirit of persecution has been shown in this matter." [8]

Chisum attended McSween's trial and his presence added interest to the case. His attendance had not been requested nor his property endangered. Chisum was there only as a friend. After the verdict of not guilty was read, the McSween friends in Lincoln rejoiced.

Shortly afterward, a mass meeting of citizens gathered in the courtroom. The object of the meeting was to discuss the violence in Lincoln County. Speeches were made about the trouble and how to end it. A committee composed of Chisum, Patron and Clenny presented several resolutions which were unanimously adopted. The first one read as follows: "That it is the sense of this meeting that our present troubles are only a continuance of old feuds, dating back for five or six years, that will now cease as the cause has been removed."

One resolution condemned the conduct of Territorial Governor S.B. Axtell. Due to his refusal to investigate their troubles, they stamped him as a little one-sided partisan and blamed him for the loss of life in the county.

Another resolution praised the good feeling that bound all the citizens together, Mexicans and Americans. "We recognize our mutual dependence upon each other and the maintenance of the laws," it read.[9]

McSween's exoneration caused only a hum of excitement generally however, since the entire country was excited over the murders of Sheriff Brady and Deputy Hindman. These two murders "fanned the flickering coals of antagonism."[10] Trials were postponed for those implicated in the death of Tunstall, and some persons were released under bond.

Marion Turner reviewed the situation in a letter to the *Las Vegas Gazette*, April 28, 1878.

> My impression is that there was a power (pecuniary) on the Rio Pecos, the prerequisites of which both parties sought to procure . . . My firm belief is that although the adherents of these parties have been guilty of killing their enemies, there was no murder in the matter, but a contest for the best of the fight, any good man would try to get.

John Chisum was blamed by many for the trouble along the Pecos. Was he to blame? Was he a man of mystery and intrigue? It is difficult to establish whether Chisum was a man of his word who paid his honest debts, or whether he found it easy to give way to the lure of larceny which is said to exist in the hearts of most men. What kind of deal did he really make with Hunter, Evans and Company? Were they to pay all of his debts? Did he have a bank account in St. Louis where money was deposited for cattle sales? Did he keep big sums of money in his safe on the ranch? He was always well funded and was considered a man of wealth. He continued to buy purebred stock as the following quote substantiates:

> In the spring of 1883 Chisum imported forty-two head of shorthorn Durhams from Missouri, reported at the time to be 'as fine a bunch of animals as ever held down hoofs.' They cost from $150 to $1500 per head and were delivered by rail to Las Vegas and driven from there to the ranch.[11]

Chisum built a frontier mansion. He traveled about on business and for his health. What verbal agreement did he have with his brothers, Pittser and James? Even if he transferred his property to them after the Hunter, Evans and Company contract matured, Chisum continued to have the say-so in everything pertaining to the ranch management. Fulton wrote that Chisum transferred most of his property to his brothers to fight the injunction of the old packinghouse notes, but did not make Pittser and James his partners.[12]

As long as he lived, after the packinghouse suits were filed, Chisum probably never had any stock or property in his own name, having first transferred his herds to Hunter, Evans and Company, and later to his brothers. Yet newspapers continued to publish the sale of John Chisum herds and his purchase of purebred stock, proof that he still ruled the ranch.

According to Harwood Hinton, editor of *Arizona and the West*, whose Master of Arts thesis was on John Chisum, Chisum never really owned a head of cattle the rest of his life after transferring his herds to Hunter, Evans and Company.[13]

A court decree dated November 14, 1884, stated that certain property previously held by his older brother, would be turned over to Pittser "so as to defraud, hinder and delay — settlement to his creditors who from that time until his death were unable to collect as he had no apparent assets."[14]

Perhaps the most dramatic example of Chisum's contradictory and at times elusive character, can be demonstrated by the manner in which he acquired the herd of cattle which he gathered up on the West Texas range and delivered to Charles Goodnight in the early days of his New Mexico ranching period. Goodnight refused to accept the cattle with their many brands, believing them stolen. Although Chisum had bills of sale from the owners and won his suit in the New Mexico court against the charge of cattle theft, were the cattle ever paid for?

If the story is true about Hunter riding into Texas and gathering up the old notes Chisum had paid for the cattle, and later paying Chisum with these old notes at par for the cattle he had bought from him, then Chisum had not paid West Texas ranchmen for their stock. There are still unanswered questions about John Chisum.

Lily Klasner wrote in her book, *My Girlhood Among Outlaws*, that Chisum planned to retire from ranching after selling most of his cattle to Hunter, Evans and Company, and would then pay for cattle which he had taken charge of to dispose along with his own.[15]

If Hunter really did pay Chisum for his cattle with Chisum's own notes, it was no doubt the most bitter disappointment Chisum ever suffered in his long experience in the cattle industry. As Haley wrote, "In his later years the fortunes of war levied heavily upon Chisum. His business suffered from thieves within and without, and, broken in health, he saw the impairment of his estate, the crowding of his range and the loss of his power." [16]

Despite his many reverses, Chisum could still force a laugh at himself. One time during his latter days, Chisum was preparing a meal in the absence of the cook. A chuck-line rider, unfamiliar with the country, stopped by the ranch and made his way to the kitchen. Drinking a cup of black coffee, the stranger looked at Chisum, shrunken, white haired and sickly, and thinking he was the cook, said, "Working for old Chisum, eh?"

Chisum grinned and replied, "Yes, workin' for old Chisum."[17]

Chapter 12

The Siege of the McSween Home and His Death

The feud raged on. Tunstall, Sheriff Brady and Deputy Sheriff Hindman had been murdered and vengeance was sworn by their supporters.

There were two factions, the sheriff's party known to be Murphy's supporters, and the McSween crowd. Dangerous and notorious outlaws fought on both sides. Jesse Evans, one of the accused slayers of Tunstall, fought on the sheriff's side. Billy the Kid, one of the alleged killers of Sheriff Brady and Deputy Hindman, was a McSween supporter.[1]

Tunstall's body, which was hauled into Lincoln in a wagon owned by John Newcomb, was taken to McSween's home. Sixty men had gathered there to learn the details of the killing. They vowed to support McSween in his efforts to see that justice was done. McSween and his men swore to avenge Tunstall's killing, despite the war cry of the Murphy–Dolan faction, "Tunstall is gone. We must get McSween next." [2]

The *Las Vegas Gazette*, June 15, 1878, described the turmoil in Lincoln County:

> When one party or the other, gets in close quarters in Lincoln County, they send out couriers to warn the people to come in and aid in the fight. These riders are kind of press officers, and make them turn out whether they want to or not. The quiet class of citizens are getting tired of fighting other peoples' battles and are leaving their farms, stock and growing crops and are seeking other and safer localities. It is fight, fight, or die, and most of the people prefer the first. The war in Lincoln County has done that section infinite harm. Such wholesale slaughter creates terror in the mind of the good, law abiding people, and those who contemplated emigrating hither will hesitate long before venturing their lives and property in so reckless a community. Bad men, on the other hand, will seek it as an asylum and hiding place from crime.

It was July, 1878; McSween rode into Lincoln from Chisum's South Spring ranch where he had gone for advice. After talking to Chisum, he had decided to ride back to town and face the consequences, although as he testified to Special Investigator Angel a few weeks earlier, he had never carried a more formidable weapon then a penknife, and it would be an easy matter to kill him.[3] He told Angel during that interview that he had heard Murphy say that he "controlled not only courts and juries, but that he could cause the death of any person who opposed him." [4]

McSween knew he was a marked man and might meet death at any turn in the road. He knew, too, that he was incapable of leadership, but was in too deep to turn back. He could not understand how he had become involved in this horrible, bloody, vendetta. At heart he was a peaceful man and had only tried to uphold the law.[5]

During those dangerous and uncertain days, McSween, fearing for his life, made his will. He left everything to his wife, Susan. He named John Chisum executor without bond, knowing that Chisum was a true friend of both himself and his wife. "I believe that were she left without means of support, he would furnish her all that would be necessary to make her comfortable during her life if it was in his power to do so." [6]

As McSween rode into Lincoln, he was gradually joined

by forty or more loyal men who rode silently along with him.[7] When they reached Lincoln ten of the riders went with McSween to his home while others were sent to the Tunstall and McSween store and to the Ellis and Montana homes. The anti-slavery forces had made John H. Copeland, but Murphy, using his influence with Governor Samuel B. Axtell, overturned the appointment and secured the job for George Peppin. The McSween party, therefore, knew there was little hope for justice.

By "grapevine telegraph," intuition and instinct, the people knew that July 19, 1878, would be the day when death would come to Lincoln Town. The entire region waited fearfully. At last they began to ride into the village from every direction — mounted men, carrying their favorite weapons. They had crossed rivers and climbed mountains to reach Lincoln. There were riders from Mesilla, from Socorro and the farms and ranches of the Pecos. Work was at a standstill. Bawling calves, thirsty for water, and cows bogged in water holes, were abandoned to their fate. The Dolan and Riley forces, ranged against McSween and his men, would clash this fateful day.[8]

The life of Lincoln was virtually paralyzed for the three days of July 16, 17 and 18. Bullets continuously whined through the air. God-fearing folk stayed inside their homes, too afraid to venture out. One bullet hit Benjamin Robinson, a Fort Stanton soldier, slightly wounding him.[9] By July 19, no arrests had been made and no warrants served on any of the men barricaded in McSween's home. As a result, Sheriff Peppin sent a note to Commanding Officer Nathan Augustus Monroe Dudley at Fort Stanton telling him that all the men he had United States warrants for were protected in McSween's home. It was impossible to serve warrants. "If it is in your power to loan me one of your howitzers, I am of the opinion the parties for whom I have said warrants would surrender without a shot being fired." [10]

The following morning Major Dudley arrived with Negro troops and two Gatlin guns, but no howitzers. He was supposed only to see that the women and children were safe, but sympathized with Murphy's camp, and refused to stop the fight. He called the besieging party a lawfully organized posse and said he could not interfere unless requested to do so by the sheriff. Colonel Dudley persuaded Justice-of-the-Peace John B. Wilson to issue warrants for the arrest of McSween and others, charging them with intent to kill Benjamin Robinson, the injured soldier. With warrants in hand, Sheriff Peppin and his men hastily made plans to arrest the men, realizing the law would uphold them in any killing that might ensue.[11]

Some sixty of Peppin's men surrounded McSween's home. Colonel Dudley's Negro troops were stationed in front of McSween's front door. The colonel demanded the arrest of the men inside. McSween refused to surrender and read the president's order which stated that the military had no authority to interfere or assist the civil authorities. During this interlude some of the sheriff's men slipped around to the rear of the McSween home, poured oil on the roof and window sills and set the house on fire. The men inside fought flames and bullets but knew the slowly burning structure would finally force them to make a run for their lives.[12]

Mrs. McSween, who had refused to leave her husband's side, had begged Colonel Dudley to save his life, but no action had been taken.[13] McSween was "taut, nervous and shaky." Being a religious man he was unarmed and praying for survival.[14] The Kid was not whipped yet. He even played Mrs. McSween's piano, (the first one in Lincoln Town)[15] to boost his companions' morale. Looking at McSween he said, "Come on Mr. McSween, don't take it so hard." [16]

Fearful that her husband would die in the burning building, Mrs. McSween crawled on her hands and knees to the front door of her home with bullets flying above her. She somehow made her way to Colonel Dudley's camp. She asked him why he permitted his soldiers to surround her home. The colonel said he had brought his soldiers to Lincoln to protect the women and children. She asked him why he permitted Sheriff Peppin's posse to burn her home and Dudley said he was not aware that Peppin's men had set the house on fire. Telling the colonel that he was a cruel, hardhearted man to let Peppin destroy her home she vowed that one day, "if she lived, his deeds would recoil on his head." [17]

McSween and his companions were again urged to surrender by Deputies Marion Turner and Robert Beckwith. When they refused to do so, shooting resumed and McSween fell near the door of his home with nine bullets in his body. Vicente Romero, Harvey Morris and Francisco Zamora were also killed. Billy the Kid escaped in the darkness, killing Beckwith as he dashed out of the door.[18]

Salazar, who left the house with Billy the Kid, had not run a dozen steps before he was hit with three bullets, the major wound being in his side. He lay unconscious for a time. When he came to, he realized the fight was over for he could hear Murphy's men celebrating with songs, laughter and drink.[19]

Salazar dared not move for fear of being killed. One man, Andy Boyle, kicked him in his injured side to see if he was really dead and Salazar bit his tongue to keep from screaming. He heard someone tell Boyle not to waste a bullet on him, that he was already dead, and Salazar thanked his guardian angel. Finally the crowd left and after three hours of extreme pain and much loss of blood, Salazar crawled down the river and reached the home of John Otero where his sister-in-law was staying. As the doctor treated him the following morning, three Murphy men walked in, having tracked him by his blood. They started to kill him, but the doctor intervened and Salazar was saved.[20]

Mrs. McSween, stunned over the death of her husband, still found refuge in the last undestroyed room of her burning home. Then, running to the Tunstall store, she joined the Ealy family who lived in the east end of the building and was conducted by soldiers to safety outside the firing zone.[21]

There were six dead in the siege of the McSween home. They were Romero, Zamora and Crawford who were buried in the old Lincoln cemetery. Beckwith's body was taken to

Seven Rivers. McSween and Morris were buried back of the McSween store near Tunstall's grave.[22]

Frank Collinson wrote that the senior Beckwith, an Englishman who had settled his ranch at the mouth of Seven Rivers in New Mexico Territory several years before the Lincoln County War, was not involved in the feud and had urged his two sons to stay out of the conflict, warning them they would be killed. However, his only daughter was married to William A. Johnson, one of Murphy's henchmen, who influenced his brothers-in-law to join up with the Seven Rivers faction. Both men were killed before the feud was over. John was killed first in one of the skirmishes and Billy the Kid shot Bob in the Lincoln Plaza fight as he, Billy, escaped from McSween's home. The old man was brokenhearted and blamed his son-in-law for their deaths. When Johnson rode out to the ranch to see his wife, who was living in her parents' home, Beckwith asked, "Where is Bob"? He was told that he had been killed by Billy the Kid. Beckwith looked harshly at his son-in-law. "You persuaded my boys to join the Seven Rivers faction. You are responsible for their deaths. I will kill you if you are here when the sun sets below the Guadalupe Mountains."

Johnson did not take his father-in-law seriously and was still visiting with the family when the sun fell below the mountain. Just then Beckwith walked around the house with his gun cocked. "Your time is up," he said, and shot his son-in-law in cold blood. Then he walked to the store and told the men there what he had done, asking them to return with him to identify the body before it was wrapped in a wagon sheet and hauled to the cemetery for burial. Collinson wrote that Johnson's death was not investigated since killings did not cause too much excitement in Lincoln County.[23]

After the killing of McSween and the escape of the rest of the men from the burning house, Peppin and his crowd began to take possession of the spoils. They knocked down the door to the Tunstall and McSween store and helped themselves to clothing, groceries, blankets, guns, and whatever took their fancy. Some knocked down the door to the bank in the rear and helped themselves to money they found there.[24]

All bodies were left where they lay until the following day. Beckwith's body had fallen over that of McSween, and was moved, the only change that took place.[25] Sheriff Peppin officially disposed of the dead, but Mrs. McSween sent word that he was not to touch the body of her dead husband. Two faithful Negroes employed by her husband attended to his burial. Only a brief service was held due to the vandals and ruffians who were swarming over the town.[26]

A letter to Tunstall's father in England from Mrs. McSween after the murder of her husband tells the tragic story of what really happened that fateful night:

> In the evening my husband and his friends, blinded with smoke, wearied by battling the flames and the no less remorseless enemies outside; driven from room to room as the fires increased in fury till the last room was consumed, at last sallied from the house and — Oh God, my heart almost did fail me to write it — was shot down like a dog on his own threshold, and when he had fallen upon his knees, calling out, "I surrender, Oh, My God, save me. Oh, Lord, save me." And after he fell, they tore up burning boards and piled them to his body and left it to burn. Oh, how can I live through this, I hope you all pray for me. I need your sympathy so much my burden is so heavy. I am left homeless, and almost clothesless and defenseless, and with nothing but the memory of a kind and loving husband to sustain me . . .[27]

Another death resulting directly from the Lincoln County War was that of Houston J. Chapman, a Las Vegas lawyer whom Mrs. McSween had employed to help her settle her husband's affairs in Lincoln County. Chapman had written a letter to Governor Lew Wallace, October 24, 1878, calling his attention "to one person whose actions have been offensive in the extreme to a large number of the best citizens of Lincoln County, and that man is Colonel Dudley." Chapman wrote that he had facts proving that Colonel Dudley was criminally responsible for the killing of McSween, and that the colonel had threatened that if martial law were declared he would arrest Mrs. McSween and her friends at once.[28]

Chapman's dogged determination to reform Lincoln County resulted in his death. He was shot down on February 18, 1879, in front of the Lincoln post office. Eye witnesses, among them Billy the Kid, said that bullets from pistols fired by William Campbell and Jesse Evans, notorious outlaws, had killed the lawyer. Chapman was buried in the Lincoln Cemetery. Although Campbell and Evans were arrested and jailed at Fort Stanton, they later escaped. Governor Wallace offered a reward of $1,000 for their capture, but by April 1879, he gave up hope of recapturing the outlaws.[29]

Colonel Dudley was later removed from his command at Fort Stanton and was granted a military board of inquiry. The inquiry, according to one writer, "was a theatrical affair, and the colonel emerged covered with a strange mixture of white ash and tar bestowed by the board of inquiry since it exonerated him." [30]

Mrs. McSween tried to convict Colonel Dudley of arson and murder, and later added a slander suit for defamation of character, but her lawsuit never got to court.[31]

It is doubtful that McSween ever paid the Emil Fritz heirs the balance of the insurance money due them. He did not ever file in Probate Court an adequate accounting complete with receipts showing how the insurance money was spent. When she administered her husband's estate, Mrs. McSween found no trace of the money. His private papers were no doubt "blown to the four winds" after the seige and burning of his home and the looting of the store.[32]

Mrs. McSween experienced as much heartbreak during the Lincoln County war as any woman in the area. Ultimately she triumphed over her misfortunes and lived to a great old age. She later established her own ranch in the Three Rivers area from funds she received from her husband's estate. She was also given the Brewer ranch of 640 acres in payment for his note which she held, dated October 1877. She married George Barber in 1884, but divorced him in 1891.[33]

An article about Mrs. McSween which was printed in the *Commercial Advertiser*, New York, and copied September 1, 1892, by the *Old Abe Eagle*, a Lincoln publication, called the lady "the most remarkable woman of this remarkable age." The article described the home she had built with her own hands; the ranch she managed, and the silver mine she had discovered on her ranch property. "She sang charmingly, played the piano and handled a cambric needle or a water color brush as dexterously as she uses an adze or a jack plane."

In her later years, Mrs. McSween sold her Tres Rios ranch to Albert Bacon Fall and moved to White Oaks, New Mexico, where she spent the remainder of her life. She died at eighty-six years in 1931.[34]

During those feuding days many settlers hitched up their teams and left Lincoln County for safer locations. The *Las Vegas Gazette*, August 17, 1878, carried the following article:

> Six wagon loads of emigrants from North and South Spring Rivers in Lincoln County passed through Tuesday going north. They were driven out by the lawless element of the section. About twenty horses had been stolen from them. A deputy sheriff rode up with his posse and demanded that they take up arms and go with them and fight. They refused and loaded up and left the country. They left their lands, houses, standing crops, gardens and everything pertaining to comfortable homes.

A short while later the same paper carried this item:

> Hunter and Evans, Wyley, Coggin, Chisum and all the big cattlemen are leaving the infected district of Lincoln.[35]

Another item in the *Las Vegas Gazette*, August 16, 1878, reported that:

> Colonel R.D. Hunter is moving every hoof of cattle he has on the Pecos across the Staked Plains to the Panhandle. He will take them to Colorado and the remainder to Wyoming. The move is occasioned by the Lincoln County War. Mr. Coggin will also send the eight thousand cattle he bought from Hunter across to the Panhandle. In a few weeks there will not be a

hundred head of stock on the Pecos. So much for a war, a county depopulated and industry banished.

In poor health and financially ruined, Murphy died soon after the Lincoln County War. Chisum's health had begun to fail also and he too would die a few years later with a terminal illness. The old scars of battle did not heal for a few years. During those fading days of the great feud, Chisum's steers continued to be shot down just for vengeance.[36] As one writer said, "Cowboys dry gulched in lonely canyons and on flats of the great wild range, and Billy the Kid, a small slope-shouldered figure rode through it all as deadly as a sidewinder rattlesnake." [37]

And what of the Hunters who were in residence at Bosque Grande in 1878 as the Lincoln County War built to its climax? The Hunters were beginning to fear for their own safety. David, his wife and sister were determined to leave as soon as the Hunter cattle could be rounded up and driven off the Chisum range. John Chisum carried a heavy mental burden. He had told them about his good friends the McSweens, and one time the Hunters had met Mrs. McSween when she visited the Chisum ranch. They had thought her a charming and lovely lady. Mrs. McSween had been stylishly dressed and was very friendly. She had driven to Spring River to talk with Chisum about the trouble in Lincoln.[38]

Hearing later about the death of Tunstall and still later about the siege of McSween's home and his death, the Hunters counted the days until they could leave the war-torn cattle range. Both factions of the feud had insisted that David Hunter join their group but he had refused, saying he had come to the Pecos to gather cattle, not to fight. Some leaders threatened his life if he did not join. These threats hastened Hunter's desire to leave as soon as possible.

John Chisum was so upset over the death of McSween and threats made against his own life, that he too decided to leave the Spring River Ranch and move his smaller herd over into Texas. Leaving Pittser and James in charge, he left one morning in a buckboard. He stopped to say goodbye to the

Hunter women at Bosque Grande and ate a hurried breakfast with them. He asked them to prepare him a lunch. As he left he said to Margaret, "God bless you Maggie, honey, you're the best child I ever raised." They never saw him again.[39]

"Soon after John Chisum told the Hunter women goodbye and drove off, two men rode up to the house armed with rifles and six-shooters asking for him," Mrs. Hunter wrote in her journal. "I told them he had been there but had gone." The men rode around asking the cook and others where Chisum was, then hitched their horses and came into the ranch house. They ate dinner there and told the women about the firing of the McSween home, how many shots were fired into his body, the slaying of the others, and how they had riddled the McSween's home with gunfire as it burned to the ground. "I wanted to scream but was too faint to move or to say a word," Mrs. Hunter wrote. "Then the men thanked us for the meal, mounted their horses and galloped away, out of our lives forever," she concluded.[40]

It was not known whether the men were friends or enemies of Chisum, and Pittser too was worried after hearing about their visit. Had they come to kill his brother? Pittser's usually cheerful disposition had vanished. John, who always told them what to do, was gone, and too many outlaws were loafing about the ranch. Sallie and her father, James, tried to keep the home fires burning but all of them realized the seriousness of conditions, and made hurried plans to move over into Texas. Frank, the faithful Negro cowboy, was also grim. He grieved for "Mistah Chisum." His small, dried up figure seemed to shrink still more. When "Mistah Chisum" was in trouble, the rest of them were also in trouble.[41]

David Hunter was also having trouble with some of his hired men, especially the rough, tough riders and hard drinkers who itched for adventure and excitement. Many of them had joined up with either the McSween or Murphy faction.[42] Jesse Evans, who had first worked with Hunter rounding up and cutting out good grade cattle with Chisum's men, had joined the Murphy side.[43]

Mrs. Hunter wrote that strange riders, crossing the Pecos by night and day, rode up to the Chisum ranch home at all hours, asking about certain parties. "I tried to be brave but my knees knocked together and my heart was in my mouth when I talked to them," she wrote.[44]

David and his remaining cowboys finally rounded up the last of the Hunter and Evans cattle and he and his wife, Margaret and sister Lizzie, left the troubled Pecos Valley August 3, 1878, heading for Trinidad, Colorado. The Chisums, too, were about ready to move over into Texas.[45]

* * * * * *

Since Colonel R.D. Hunter played such a major part in John Chisum's life, it is appropriate here to provide some additional background about his career.

The colonel helped promote, and served as chairman of the first convention of cattlemen in the United States in St. Louis, Missouri, in 1884, and was later elected the first president of the Association. Hunter and Evans at that time owned several ranches in Texas, Indian Territory, Kansas, and Nebraska and had 386,000 cattle and a capital of $5,620,000. They owned otherwise or controlled by lease 11,464,000 acres of land. The blizzard of 1886–87 was disastrous to them and they suffered staggering losses. The company dissolved in 1890. Colonel Hunter came to Texas and with other investors bought a coal mine in Thurber, Texas, which was called Texas and Pacific Coal Company. William W. Johnson and his brother, Harvey, sunk the first shaft on their ranch and operated the mine a short while before selling it to the Texas and Pacific Coal Company. Colonel Hunter headed the company as president. He moved his family to Fort Worth and built a palatial home on Summit Avenue. He died in that city in 1902.[46]

Chapter 13

Removal to Texas: Governor Wallace Comes to New Mexico

Chisum was in St. Louis, Missouri, receiving treatment for an old leg injury when McSween was killed.[1] One can but sense his grief and concern upon returning home and learning of the five-day siege of Lincoln Town, the burning of McSween's home and the cold-blooded gunning down of the unarmed McSween at his own front door. No doubt, Mrs. McSween poured out her grief to her good friend, Chisum.

The country was in such a dangerous mood and so much turmoil that Chisum made hasty plans to abandon his ranch. He moved his small select herd into Texas to the Canadian River country. Trying to ensure the safety of his family weighed him down. He realized that conditions would continue to worsen and that he and his family were not safe on the ranch.

An earlier event which occurred prior to the death of McSween worried the Chisum family and contributed to their growing insecurity. A Chisum cowboy named McNab, a sympathizer of McSween, started a rumor that a movement was under way to drive all of the rustlers out of the Seven Rivers country. McNab and his men were ambushed by Sheriff Peppin near Lincoln. McNab was killed and his men captured. On their way back to Lincoln, Peppin and his posse saw a lone horseman in the distance and pursued him. The rider was James Chisum, who outdistanced his pursuers after a twelve-mile chase and returned safely to the ranch. When James recounted the story to his brothers, John realized his family was really in danger.[2]

John next heard of a vicious plot to disfigure his brother, Pittser. At the time Pittser was visiting at the home of Captain Lea in Roswell. His brother, James, was on the ranch and John Chisum was away from home. Some of Chisum's enemies planned to waylay and to mutilate Pittser on his return to the ranch. They knew he was visiting in the Lea home. They planned to brand him with a Long Rail and give him a jinglebob ear mark. They were drinking behind Captain Lea's home and boasting of their plan when an outsider learned of the proposed attack and warned Lea. The captain sent a Negro boy to saddle Pittser's horse. The boy led the mount to the front gate and Pittser was hurried off and was safely home before his escape was known.[3]

As was stated earlier, Chisum left home ahead of his family in late July, 1878, traveling in a buckboard. He stopped to say goodbye to the Hunter women at Bosque Grande. They understood he was heading for Las Vegas, New Mexico. It is not known where John Chisum spent the next few weeks but he was doubtless looking after business affairs. He was in Las Vegas during the middle of August because Sallie mentioned in her journal a letter from her Uncle John which had been mailed from that town. Chisum wrote that he had made arrangements to send her and her brothers to school in Anton Chico, New Mexico, that fall.[4]

Soon Sallie, too, was on her way to Texas with her father and two brothers, Will and Walter. The several thousand heifers that John Chisum had kept as the nucleus for his select herd had been gathered up and headed for Texas and James

Chisum and his three children were part of the accompanying entourage. Pittser Chisum had elected to remain at South Spring Ranch headquarters with a skeleton crew.

The trail caravan stopped for several weeks at Bosque Grande. Charles Nebow was in charge of the drive and Negro Frank was his right-hand man. Other cowboys were Johnny Newell, Bill Hatch and Tom Picket.[5] There were two wagons loaded with baggage. Sallie drove one of the teams. There were also 150 horses herded along with the cattle. When they departed Bosque Grande, they moved up the east bank of the Pecos, stopping at Fort Sumner where they were joined by John Chisum.[6]

The caravan turned north ten miles east of Sumner, skirting along the eastern edge of the Taiban mesa. After passing Fort Bascom, near present Tucumcari, they reached the southern drainage of the Canadian river. The last camp in New Mexico was at Ute Creek after which the Texas line would be crossed September 28, 1878.[7]

After moving the herds into the Canadian River country, the Chisums lived in temporary dugout camps with corrals at the foot of the breaks near the river below the mouth of Turjilla Creek. Their camp was about thirty miles from the frontier town of Tascosa where supplies could be bought. John's pretty niece, Sallie, added the only feminine touch to the cow camp.[8]

Frank Collinson, the English cowboy, wrote that the great buffalo hunt was over in 1878 and he and other buffalo hunters were camped in Blanco Canyon in the Llano Estacado, Crosby County, Texas, near the rock house of Uncle Frank Smith.[9] Some of the men planned to go west to work on the railroads. Collinson wanted to go over into New Mexico to fight in the Lincoln County War where it was rumored a man who knew how to use a gun could earn four dollars a day. After supper, as the men sat talking around a camp fire, a hack pulled up to the Frank Smith home. The men in the vehicle were Pittser Chisum and S.R. Coggin. Coggin was looking for a Texas location for 8,000 jinglebob cattle he and

Wylie had received from John Chisum in payment of a debt. No doubt Pittser was looking for a suitable range for Chisum cattle.[10]

The two men joined the buffalo hunters that night and Coggin, who knew Collinson, asked him if he knew the country well. Collinson told him that he knew it "from Dodge City to the Concho." Coggin then asked Collinson if he would go with him to Fort Sumner and drive the herd over into Texas to a good range. Collinson hesitated. "I had planned to go to Lincoln County to fight in the war where an experienced sharpshooter can make good wages," he said.[11]

Coggin shook his head. "You would be killed," he said, "and the rumor is not true anyway. Pittser will tell you that his brother John is not offering wages. In fact John has chosen to remove himself and his affairs from the battlefield by going to Texas." [12]

Collinson then agreed to accept Coggin's proposition. He would go with him to Fort Sumner, drive the herd over into Texas to a good range. He knew the country he would choose. It would be the old buffalo hunting ground on Tongue River. Collinson and Coggin started back to the Pecos the following morning. Pittser remained in Texas. No doubt he scouted the country and chose the location for the Chisum camp, and the surrounding range for their cattle.

Collinson wrote that he counted the Coggins and Wylie cattle as they were classed and shaped up for their new owners. A St. Louis bookkeeper counted for Chisum. By the middle of August the herds were on the trail to Texas. Collinson piloted the young stuff, firsts, seconds and three-year-old heifers and steers. These cattle, according to Collinson, were the first herds to be driven into Texas from Fort Sumner over the old Mexican buffalo hunters' trail.[13]

Collinson reminisced about the dramatic times on the Pecos when he arrived there with Coggin. During the roundup of the cattle he spent a good deal of time around Fort Sumner and it was there that he first saw Billy the Kid. During this time it was rumored that the Seven Rivers outfit planned to attack South Spring Ranch. All hands were ordered to get there as quickly as possible. Collinson ran to the wagon and took out his Sharps .45 where it was kept in a blanket. He joined the other cowboys as they raced toward Chisum's headquarters. They found the Kid and his gang there and the Seven Rivers crowd had cleared out in a trail of dust. Collinson said he was disappointed that there was no gunfight, but it took place later in the McSween home.[14]

R.K. Wylie piloted the cows and calves belonging to himself and Coggin into Texas. Due to the shortage of water and the long drive, about one hundred head were lost but they finally passed through Blanco Canyon and the following day reached the Tongue River. That range was all open country. Water, grass, and wood were plentiful. Deer were in every thicket and antelopes on every flat. Plums, currants and grapes were in abundance. "Yes where on earth was there better natural pasture with all the brakes running east, the Cap Rock on the west and the finest water in Texas. I have ridden horseback over most of the West, from Old Mexico to Canada," he wrote, "but never have I seen its equal as it was in those days — no wonder the old Comanche buffalo-eating Indian hated to give it up." [15]

Collinson established camps at Dutchman Creek and his home was a dugout used by a buffalo hunter. Corrals were finished, the jinglebob cattle, formerly owned by Chisum, waxed fat on the virgin range. This was the beginning of the famous Matador Ranch.[16]

In the contract with Hunter, Evans and Company, $27,000 worth of stock was to be turned over to Pittser, John Chisum's brother, when the deal was closed.[17] During the Chisum stay in the Panhandle of Texas the deal was terminated and the St. Louis firm turned over cattle valued at this amount to Pittser. Title to John Chisum's registered herd was also in the name of his brothers. The letter "U" — high up at the tip of the left shoulder — had been chosen as their new brand. It became well-known and was recorded in the brand books of San Miguel County, New Mexico, 1879.[18] When the brand was at its zenith a reporter wrote the following after riding over the Chisum range:

> U will know John's brand wherever U see it and if U should happen to visit the lower country U will see it on about 20 to 30 thousand head.[19]

History was being made over in New Mexico while the Chisums bided their time on the Canadian, counting the days until they could move back to the comfortable South Spring Ranch home.

Indirectly Chisum was responsible for law and order coming to the New Mexico Territory. Dr. Montague R. Leverson, a visitor at the Chisum Ranch before the move to Texas, was looking for a suitable place to establish settlers. He was a friend of President Rutherford B. Hayes. He just happened to arrive in Lincoln County at the height of the feud and realized that law was sadly lacking. He heard Chisum's side of the trouble and believed him. He immediately wrote two letters, one to Carl Schurz, Secretary of the Interior, and the other to the president. He told them facts about lawless conditions in Lincoln County and wrote that the governor of the territory "was the mainstay of thieves and murderers." He told how Chisum had been refused aid by the sheriff, by the military and by the governor to protect his cattle from bandits, who in bands of twenty were stealing cattle from his employers, Hunter, Evans and Company.[20] He wrote that all United States officers except the marshal and his deputies were at the head of "a gang of veritable thieves."[21] He wrote President Hayes that he had come to New Mexico to find suitable land along the Pecos for a colony from England, but unable to find the security required, would have to look elsewhere.[22]

Before closing his letter to the president, Leverson reminded him that back in 1868 he had "done the nation some service by uncovering certain widespread naturalization frauds. He had boldly exposed the frauds over his own signature in the *New York Tribune* and had refused a $3,000 bribe to keep quiet. He urged President Hayes to investigate the Lincoln County situation.

After receiving the Leverson letters, the United States government began to show more interest in Lincoln County. The British embassy was investigating the killing of Tunstall and letters from McSween (before his death) and others were calling the attention of the government to fraud in Indian affairs.[23]

Apparently Leverson's letter carried weight with the President of the United States. Samuel B. Axtell was dismissed and Lew Wallace, soldier and novelist, replaced him in 1878 as governor of New Mexico. It was a strange combination of jobs that this unusual man had undertaken. It was his duty to bring law and order to the Territory and put an end to the most famous feud in the Southwest, the Lincoln County War. During the day the governor dealt with all types of men, murderers, rustlers, and listened to continuous tales of woe and crime. In the evening he retired to his office and became the willing slave of his aesthetic imagination as he wrote long into the night on his famous novel, *Ben Hur*, which was finished in 1880.[24] He soon realized, however, that he would need national help in bringing peace to the Territory. He wrote President Hayes telling him about the trouble: the United States Marshal had warrants unserved, courts were inoperative due to partisan feelings of all the citizenry, and an obvious bias on the part of the executors of the law was evident. He suggested that President Hayes declare a state of insurrection existent so that troops could be ordered into the field. The president executed such an order and on November 14, 1878, Governor Wallace issued a general amnesty to all except those charged with crimes, if they would lay down their arms.[25]

Although John Chisum and others with interests in the lower Pecos called upon the new head of state and suggested plans to quiet the unrest, they had little impact on the governor. Chisum at his camp on the Canadian realized that the governor's amnesty program would do little to guarantee safety for man or beast and that old hatreds and animosities would not be soon forgotten.

After the military purge in Lincoln County, the Chisums returned to their South Spring headquarters and by November 1879, were once again installed in the comfortable house that they loved.[26] That same month four detachments of cavalry began to inspect different ranches for stolen stock. Company commanders were furnished lists of all legitimate brands registered in Lincoln County along with descriptions of some thirty-five men wanted for murder. All camps were to be inspected and stolen stock returned to the county seat.[27]

Pittser Chisum wrote the commanding officer at Fort Stanton reporting peculiarly marked cattle in the Bosque Grande vicinity and said that persons giving these brands had never owned cattle until recently.

John Chisum wrote the governor a letter from Fort Sumner April 5, 1879, suggesting a safe way to protect the country, but his letter, which follows, had little effect:

> If ten good men were stationed at a large spring twenty miles east of this place (Fort Sumner) it would prevent robbers from coming in off the plains on to the Pecos and give protection to this place and the citizens below. Roswell is so far below that the robbers can come in and rob as low down as Bosque Grande, a distance of fifty miles, and get out unmolested before the troops from Roswell can reach them. If twenty men could be stationed at Pope's Crossing on the Pecos it would prevent them from coming up the Pecos. Having those two points guarded you then have possession of the main keys to the settlements on the Pecos River. Pat Garrett, who resides here, would be a very suitable man to take charge of the squad east of this place if authorized to do so.
>
> I hope Governor, you will not think I show any disposition to meddle or dictate. I know the country well, and I am satisfied you are more than anxious to give the citizens protection and I am equally anxious to see the robbers kept out, hence I make these suggestions. Robbers cannot very well reach the Pecos by any other route on account of the lack of water.[28]

District Court opened in Lincoln and Governor Wallace

centered his attention on obtaining convictions. Over 200 indictments had been made by the grand jury, but only a few received legal action. Local newspapers covered the trials. The editors accused Chisum of influencing the military to make the arrests. Unexpectedly the *Independent*, May 24, 1879, carried the following item:

> Unfortunate John Chisum! How much unrest and agony of mind you have caused the News scribes. Although far from the source of the strife, you are still remembered in their dealings.

By 1881 Chisum's ranch began to have definite boundaries due to new cattle companies moving in. It no longer reached from Bosque Grande to the New Mexico-Texas line, but was confined to about sixty miles in length. Many of the original line camps were abandoned but Yellow Lake, twenty miles northwest of Roswell; Stirrup Bend east of Artesia; Prickly Pear, seventeen miles northeast of headquarters, Buffalo Valley, twenty miles east and Good Bend, sixteen miles to the southeast, still existed and Chisum's ranch was considered one of the finest spreads in the territory.[29]

George W. Littlefield had closed out his Texas Panhandle holdings and had bought the land and buildings at Bosque Grande, formerly owned by Chisum. His ranch was called the LFD. Phelps White managed the ranch and was so impressed with Chisum's adjoining spread that he wrote a relative about it:

> I wish you could see old man Chisum's ranch and cattle, the best in the Territory. His home cost him twelve thousand dol. All well fitted out, but the old man will have to leave soon as he is getting old.[30]

Chapter 14

South Spring River Ranch— An Oasis in the Desert

Chisum's best years were spent at South Spring River Ranch. Sallie Chisum, John's niece, always said that Christmas 1880 was one of the happiest days of her life. The Chisums were home again from their Texas exile. Once again they were installed in the commodious and comfortable house they all called home. John Chisum invited his friends to a festive Christmas dinner and they came from miles around. The long table in the dining room, which easily sat twenty or more, was loaded with delicious foods, including barbecued meats, baked turkey and ham, salads, cakes and pies. An all-night dance followed and visiting cowboys did not ride away until the sun was coming up. It was a day never to be forgotten and Sallie often compared it to the first Christmas she spent on the ranch, which was indeed quite different.

On that occasion, Sallie, her father, James, and brothers, Will and Walter, arrived at South Spring weary and expectant on Christmas Eve, 1877. They had been traveling for weeks from their home in Denton County, Texas. Sallie's mother had died and her Uncle John had written to them, urging them to come to the Pecos and make their home with him. James could be of much help on the ranch and the virgin, open country was a fine place to bring up youngsters. Sallie's father accepted his brother's generous offer and followed the route John mapped out for them. Their caravan consisted of three wagons and a horse-drawn hack, about fifty head of cattle and a remuda of horses. They were met at the treacherous Horsehead Crossing by Chisum cowboys and crossed the Pecos at Pope's Butterfield stage crossing. They made it safely through the dangerous Black River country where Pittser had lost an entire herd of cattle that had been stampeded and driven off by the Indians. The Chisum ranch home was reached about eleven o'clock Christmas night, but there was no welcome from John. He was still in jail in Las Vegas, being held for debt. After a bite to eat, the tired family fell into bed after the menfolk had safely penned the stock in the corrals. Christmas Day dawned clear and bright but there was a stillness about the place that disturbed James. He arose early and went out to check the horses and cattle. The corrals were empty. The Indians had slipped up during the night, lifted the gates from their hinges and driven off all the stock. So passed the first Christmas on the Pecos River, a far cry from the big celebration that took place in 1880.[1]

The Long House, the name John had given his palatial frontier home, was the most comfortable he had ever known, and he enjoyed it to the fullest. He loved company. His dining table was usually filled with family, friends and cowboys. The Long House had cost Chisum thousands of dollars to build. It was an adobe structure of fort-like design with thick, almost impenetrable walls. It was by far the most outstanding ranch home in the Territory of New Mexico and the hospitality extended there was enjoyed by many early-day travelers.

E.A. Cahoon, Roswell banker, who visited the Chisum ranch, described it as follows: "The old Chisum ranch was such a wonderful place, very pretty, fine orchard and garden, big house (the front porch 144 feet long), beautiful alfalfa fields and best of all, such good, kind, generous people." [2]

The house faced west and was 150 feet long and sixteen feet wide. There were four rooms on each side which opened onto a ten-foot wide central hall. The house was plastered on the interior. Chisum had his office in the first room off the hallway and it contained a small safe, a walnut writing desk, a large dictionary on a stand and a comfortable bed. But the cowman did not sleep in that bed. It was there only for the sake of appearance. Each night John Chisum pulled his blankets from the closet and spread them on the floor, the bed he had preferred most of his life when not pillowing his head on the ground around a cow camp.[3]

This fine home on South Spring River was described by a visitor who said it was wonderfully modern in all its equipment and furnishings. "The fact that his home was 200 miles from a railroad had not deterred Chisum from providing a home with everything the East might have to offer."[4]

There were two flat-topped utility rooms in the rear of the building. One was used for a commissary. The other was divided between living quarters for the cook and a dance hall. It was said that Chisum would not permit the cowboys to dance on his new Axminster rug in the big house.[5]

A small canal of running water from the South Spring River coursed under the hallway of the house into the garden and was used for cooking and drinking.

Chisum, who loved fruit and shade, soon had willow and cottonwood trees planted about his home and fruit trees in a nearby orchard. Sallie said when her Uncle John planted the cottonwood trees that they were hauled from Las Vegas by mule pack-train, that three young trees were planted close together and the trunks bound together. In time, as the trees grew, these individual trees formed one massive tree. It was called the "Three Brothers" for John, Pittser and James.[6]

Chisum had several large *acequias* (canals) dug about his house and with plenty of water he soon turned his headquarters into a beautiful, naturally-cooled oasis. He was a pioneer who not only recognized the agricultural potentialities of the Pecos Valley, but brought in many new immigrants. By 1877, one large Chisum irrigation ditch watered 1000 acres. It was located near the head of South Spring River and not only watered the yard but a nearby field. Chisum experimented with alfalfa and caused that crop to become a mainstay in the area.[7] James Chisum helped with the farming as his talents were more for farming than ranching. Felix McKittrick supervised the farming on the Chisum ranch since he had experimented with the growing of wheat, buckwheat and rye on his adjoining place. McKittrick was an Irish cowboy who had worked for Chisum many years and was a favorite among all the hands. He was jolly and full of pranks and Chisum said he was so much a part of the outfit that he would have him marked and branded. McKittrick later established his own ranch and had a partner named Thomas. Their cattle must have looked almost as odd as the jinglebobs. They had no horns, nor ears, and were branded with a "6" on the shoulder, on the side and on the hip. McKittrick, who drank too much, later drowned in shallow water, "too drunk to lift himself out." [8] A canyon on the south of the Guadalupe Mountains carries McKittrick's name today.[9]

When new farmers moved into the Pecos Valley they were welcomed by Chisum. He gave them permission to farm along his irrigation ditches. The large area east of his home became known as The Farms.[10] On one occasion Chisum sent an ox wagon one hundred miles to the Davis Mountains for cottonwood and willow saplings to line his own irrigation ditches as well as those of his neighbors.[11] The cottonwood trees were hauled by mule packtrain from Las Vegas and planted around his home and along a driveway from the road to the house. Fruit trees — apple, peach and pear — were brought from Arkansas to plant an orchard.[12] Chisum has been credited with introducing the popular bobwhite quail into New Mexico and planting the first rose bushes in the yard of his ranch home.

The editor of the *Las Vegas Gazetter*, August 6, 1880, published a letter received from Chisum:

> You will find, in a box sent you, two heads of millet I raised on my ranch. I have been trying the different kinds of grasses and millet to find which suits our climate and soil best. From all I have been able to learn, this I send you is the best. It grows from four to six feet high, producing a large quantity of fodder and making more feed to the acre than any other feed. If what I send you proves to be the best I must be content.

The editor added a note below the letter which read, "The heads of the millet are larger than ears of popcorn and are unusually well filled."

Chisum pioneered the cultivation of New Mexico apples. He began the experiment in a five-acre orchard on the South Spring River Ranch. The result was a large, luscious apple. Today this section of the state between Roswell and Haverman is famous for its apples.[13]

Chisum was unusually helpful to Mormon settlers who came to the Pecos. He was a good neighbor, lending them farming equipment and furnishing them provisions from his store. He helped them with their building. It was reported that seven new families who had moved into the Chisum territory from Seven Rivers were encouraged to do so by Chisum.[14]

Little wonder that Sallie Chisum loved her uncle's innovative and eventful life style and cherished the memories of that experience as long as she lived. She was only twenty when she arrived at the ranch and was a favorite of the cowboys, turning the heads of many of them, including Billy the Kid. Sallie acted as hostess in her uncle's hospitable home for several years, meeting and welcoming visitors, both famous and ordinary, to the ranch.

When Sallie first came to the ranch the cowboys were timid about making friendly advances to her, but they all secretly admired her and claimed her for their sweetheart. Soon she was being called "Queen of the Jinglebob Ranch." [15] Sallie could have had her pick of any of the cowboys on the ranch, but she chose William Robert, the young bookkeeper of her uncle's ranching affairs whom she later married.

Sallie knew her Uncle John well and understood his feeling regarding the fairer sex. She said her uncle was too busy to get married, but he had told her on one occasion that he would marry and settle down when his herd increased to 100,000.[16]

It was rumored around the cow camps that Chisum had proposed to Mrs. McSween, the pretty widow of Chisum's lawyer and friend. A year after her husband was killed Chisum rounded up 200 choice jinglebob heifers and turned them over to Mrs. McSween to pay for services her husband had rendered him.[17]

When asked if her Uncle John had fallen in love with Mrs. McSween, Sallie thought that perhaps he had, since the young widow was beautiful, talented and intelligent. "It took such people to attract the attention of Uncle John," she said. "I remember that my uncle put on his best bib and tucker when driving the cattle to Mrs. McSween's home. He wore his best handmade boots and Stetson hat." [18]

Sallie also recalled that when Chisum delivered the cattle to Mrs. McSween the attractive widow asked him if the jinglebob earmark could be used on the increase of the herd. Chisum was supposed to have replied, "Not unless you give me the promise that some time in the near future we might merge the herds and you become the queen of the jinglebobs." But about that time a jinglebob cowboy dashed up on his pony and told Chisum that he was on the trail for stolen cattle and "Uncle John turned his horse around and joined him," Sallie said.[19]

In contrast to Sallie's views, Georgia B. Redfield wrote in her article "The Historical Chisum Trail," *Frontier Times*, December 1945, that one essential characteristic — the love angle — was woefully lacking in the makeup of John Chisum, the colorful hero of the cattlemen's world. "As far as one can learn from relatives and friends," Mrs. Redfield wrote, "Chisum never looked a second time on any woman with the desire of making her his wife. He never married, and there was no hint of the swashbuckling hero or lover in his quiet, unassuming, but self-assured demeanor." [20] Mrs. Redfield was apparently not aware of the more romantic image Sallie attributed to her Uncle John.

Sophie Poe, who visited on the Chisum Ranch in her girlhood wrote in her book, *Buckboard Days* how she felt when first meeting "Uncle John" Chisum, her favorite name for the hospitable cowman:

> Somehow I liked him instantly, as he held my small hand in a vicelike [*sic*, viselike] grip. I realized his strong, frank personality. His dignity showed him to be a man of importance and his genial laugh showed that he knew how to be human. He was a self-made man who had but little opportunity for schooling, yet he had remarkable business ability and his opinions were valued by everyone.[21]

Chisum teased Sophie during her visit, facetiously calling her Mrs. Chisum as he ushered her into the dining room for dinner. Before she left he gave her a beautiful pair of gold bracelets which he had ordered from the east.[22]

Chisum was often teased about his single state but laughed it off, saying "the girl did not court me enough."[23] Another woman friend in whose home Chisum often visited said that he loved money, was dictatorial and accused him of cutting corners when it was to his advantage.[24]

In her book, *My Girlhood Among Outlaws*, Lily Klasner wrote about her visits to the Chisum ranch and the happy days spent there. "Chisum had a flirtatious reputation with several young women who visited the ranch, but most of them understood his attentions," she wrote.[25]

Chisum's loud belly laugh and the jovial personality of his more youthful days have been described by several writers. Could that have been forced merriment to hide his many misfortunes? One friend said that no one laughed as easily or as heartily as John Chisum. "You would think he never had a care or a sorrow in his life, but like everybody else he had them. However, it was his way to laugh them away." [26]

The average man would find it hard to laugh after experiencing the many reverses that John Chisum had in his lifetime. He lost entire herds of cattle; was said to have been paid for other herds with his own defunct notes at par; was harassed and held in jail over unpaid packinghouse notes and was disappointed in love. Disillusioned, he shied away from marriage and remained a bachelor.

Lily Klasner's father and Chisum were old friends, had known each other in Texas, and later were neighbors in Lincoln County. Chisum often spent a night in their home. Lily was a frequent guest in his ranch home through the years, and held many of his confidences. Chisum told her about a young girl whom he had fallen in love with in Texas during his younger days. He had proposed to her, but uncertain then of Chisum's financial future, she promised to marry him if he would wait a year. Chisum did not want to wait that long and their romance cooled. The young lady later married "a city dude" and her marriage turned out unhappily. Chisum grieved about her and wished he had waited the year she had requested. He later became wealthy while she lived in poverty.[27]

Did John Chisum ever think of Jensie, the little slave girl and the two daughters he fathered by her in the years that followed? Undoubtedly so, though interracial alliances of that kind at that time were socially unacceptable. Jensie's image and memory may have been Chisum's reason for never marrying.

There is an historical marker near Chisum's old Bolivar, Texas ranch that tells the story of his pioneer ranching days. Before he died, Emory Jones, a grandson, said the marker was about the only thing the family had left of John Chisum. How wrong he was. Today there are over ninety Chisum descendants residing in Texas and elsewhere and these do not include the descendants of Harriet, eldest child of Chisum and Jensie, who married when young and moved to another part of the state.

John Chisum descendants are teachers, musicians and

businessmen in the mainstream of life. For instance, Mrs. Betty Jean Jones Foreman, only child of Mr. and Mrs. Jinks Jones, has her Bachelor of Science Degree and her Master's Degree in Elementary Education, and teaches in the Houston, Texas, Public School System.

The three sons of Mr. and Mrs. Emory Jones are all college graduates with two or more degrees each. Emory Odell Jones, who lives in Columbia, Missouri, is with the United States Department of Agriculture, Meat and Poultry division. Bobby Ray Jones is a major in the Air Force, stationed in Philadelphia. William LaRue Jones resides in Minneapolis and is Director and Administrator of the Greater Twin Cities Youth Symphonies.

Other Chisum descendants have made their mark in the world. All are proud of their heritage and never tire of the love story of Jensie the little slave girl and her cattle baron. Prized possessions in the homes of the grandchildren are enlarged photographs of their grandfather.[28]

Chapter 15

Billy The Kid

Billy the Kid had been a welcome guest on the Chisum Ranch before he became an outlaw. He and Sallie, Chisum's niece, were good friends. Sallie kept a journal and several fading penciled notes apply to Billy.

> Indian tobacco sack presented to me on the 13th of August by William Bonny [*sic*] and two candi [*sic*] hearts given me by William Bonney on the 22nd of August.[1]

Billy respected Chisum in those early days, too, but his admiration for the Jinglebob King turned to hatred when Chisum refused to pay him for fighting in the Lincoln County War. Whether there had ever been financial agreement between Chisum and Billy is not known, but Billy seemed to think there was and later made it his business to wage war on Chisum, stealing jinglebob cattle at every opportunity, driving them to Three Rivers and selling them to Pat Coughlin.[2]

Back in the early 1930s, Eugene Cunningham, a writer, interviewed Florenzio Chaves, then an old man, who had fought with Billy the Kid. Chaves told Cunningham that Billy never worked for John Chisum. According to Chaves, Billy worked a short while for Frank Poe on his Ruidoso ranch and was working for Tunstall on the Rio Feliz when Tunstall was murdered by the drunken posse.[3]

Whether Billy was ever actually on the Chisum payroll is a moot question. Some say he was, some he wasn't. Margaret Hunter, who spent two years in the Pecos Valley while her husband David gathered up the Chisum cattle purchased by Hunter, Evans and Company, wrote in her journal that Billy worked for Chisum and that they saw him often about the ranch.[4]

According to contemporaries, Billy was not a romantic figure. Frank Collinson wrote that Billy was about eighteen years old when he met him, but looked older. "He was sunburned and not much to look at. He didn't look like a desperado. If I had seen him somewhere else I would not have looked at him twice. There were scores just like him up and down the Pecos. All the clothes Billy wore would not have cost over five dollars, including an old black slouch hat, well worn pants and scuffed boots. He wore a black cotton handkerchief about his neck, old shirt and vest. He had large front teeth, no chin, narrow shoulders, small hands and feet and weighed about 135 pounds. His eyes were gray-blue and were constantly looking around." [5]

Collinson saw Billy the Kid the last time in 1880 at Fort Sumner where Collinson judged some Shorthorn bull yearlings. Billy was accused of stealing cattle on the Canadian River and driving them to the Rio Grande. Collinson wrote that this was foolish talk because he had driven cattle over this route and it took many hands, horses, wagons and barrels of water, and even with those preparations many cattle died. "This kid, and a man or two, driving cattle over that dry territory, with chuck tied behind their saddles, and a canteen of water, could not be done," he wrote.[6]

Collinson said he always felt that the Kid was not treated fairly because he was the only killer ever brought to trial for any of the many killings in the territory from 1878 to 1880. Collinson did not believe that Billy killed for money or pay of any kind, but solely for the revenge of Tunstall. "Tunstall was the only man who ever treated me as if I were freeborn and white," Billy told Collinson.[7]

After the killing of Brady and Hindman by Billy and his warriors they lost the friendship of many of Lincoln's citizens and were looked upon as outlaws in every sense of the word. Billy had been slightly injured when picking up the dead Brady's rifle. He had been shot at by Mathews who had found refuge behind a nearby house. The ball from Mathews's gun had shattered the stock of Brady's rifle and glazed Billy's rib cage but not seriously. He quickly fled to the McSween home for protection, but he and his gang were soon riding again.[8]

Rumors were rife about Billy and his gang throughout the Territory. Law-abiding citizens did not feel safe as long as he was free. It was whispered around that he planned to kill John Chisum and it was reported that Chisum, for that reason, or for business reasons, left the ranch and for a while his whereabouts was unknown.

The story was told about Billy sitting in the sunshine in front of a saloon one day when he saw some snowbirds pecking in the sand. He said to his pals, "Suppose that pretty little bird was old John Chisum and if I was to shoot and hit him anywhere except in the head it would be murder." He fired and a bystander picked up the dead bird. Its head had been shot off. "No matter," Billy cried, "let's give old John another chance." Another bird's head disappeared. After killing several in this manner Billy hit one in the breast. "I've murdered old John at last," he said. "Let's go get a drink." [9]

Stories about Billy the Kid are as much legend as truth, but it was a fact that his friendship for Chisum turned into hate and vengefulness after the Lincoln County War. Billy seemed to think Chisum owed him money for protecting his range and for fighting on the McSween side. On the other hand, this story might have been his excuse for stealing jinglebob cattle.

Several writers have told the story of how Billy and Chisum met in a saloon. Billy demanded payment for the debt he claimed Chisum owed him. It is entertaining to read the different versions of this story. George B. Anderson in his *History of New Mexico* tells the story this way:

> Billy drew his gun one day on Chisum in Fort Sumner and demanded payment for service rendered. Chisum coolly walked out of the saloon, not saying a word. It was said that Billy later told one of his friends that Chisum's coolness saved his life.[10]

Maurice Garland Fulton wrote in his book, *The Lincoln County War*, that Chisum and Billy met in Fort Sumner and Billy demanded payment of $500 for services to the Tunstall–McSween–Chisum party's interests. He had not received any money since Tunstall's death, but still served as he seemed to think, under sponsorship of Chisum. He had also helped to eliminate some of the old Jesse Evans gang which had been stealing Chisum's cattle. Although threatened with Billy's pistol, Chisum refused to pay and remained calm with the muzzle of Billy's gun in his face. Reaching into his pocket for his pipe and tobacco, he said, "Billy, let me get a smoke for then I can talk better. Now, Billy, listen. If you talked about that money until your hair was as white as mine you could not convince me that I owed it to you. Billy, you couldn't shoot an honest man, could you, while he was looking you square in the eye? You have killed several men, I know, but they needed killing."

Chisum then reminded Billy that during the Lincoln County War that McSween's party had hidden on his ranch and he had provided them with whatever they needed from his store. Billy then lowered his pistol saying, "All right, Mr. Chisum, but only for the present. If you won't pay me that $500 in money I'll steal from your cattle until I get it." [11]

For the time being Chisum, with his cool composure, had pacified the Kid, but there was a grudge in Billy's heart. During the spring of 1881 he rounded up new recruits, one of them Dave Rudabaugh, who was wanted for murder. Soon he and his gang were raiding ranches, the scourge of Texas cowmen.[12]

Most of these legends about Billy the Kid and John Chisum were said to take place in a saloon and usually ended with Billy and Chisum taking a drink together. According to Doctor Henry F. Hoyt, a pioneer Panhandle Texas citizen, who knew Billy well, the young outlaw was a teetotaler, as he, himself was. One wonders then how many of the stories told about Billy during the past century were true or legendary. It is remarkable that his memory is as much alive today as it was a century ago. His life has become a part of the permanent folklore of the frontier.

After the Lincoln County War, Billy and his followers stole 125 horses from the Seven Rivers Warriors who had fought on the Murphy side and drove them into the Panhandle of Texas where there was no law and outlaws ruled supreme.[13] Dr. Hoyt was the first doctor in the area and served later as Surgeon General of the United States Army in the Philippines.[14] He told about meeting Billy and his companions when they rode into Texas. Hoyt, who was substituting as mail carrier during the illness of the regular carrier, was returning on horseback to Tascosa when he met a party of five strangers riding toward the town. They turned out to be Billy the Kid, Tom O'Folliard, known as Big Foot Tom, Henry Brown, Fred Waite and John Middleton, a known outlaw.

Late in October 1877, Chisum and his cowboys were driving cattle near Roswell when they met Doctor Henry Hoyt and Hugh McCane. The two young men had been told that Chisum might hire them. They camped with the Chisum cowboys that night, and around the campfire they mentioned the subject of work to Chisum. Chisum pointed to the east saying, "Over there is the Texas Panhandle, a big country, full of people. An epidemic of smallpox and no doctors. There's the place you're looking for." This is how it happened that Dr. Hoyt was in Tascosa when Billy the Kid rode into that frontier town and the two became friends.[15]

Dr. Hoyt wrote in his book, *A Frontier Doctor*, that the cowmen in the country knew all about Billy and his part in the much talked-about Lincoln County War. They knew there was a reward for Billy, dead or alive, but they told him he would be left alone if he behaved himself.[16]

For a while Billy and his friends lived recklessly in old Tascosa. They were not disturbed by the law. They became a part of the town, gambling, racing horses, shooting at targets, drinking (except Billy) and whoring. Temple Houston, a son of Sam Houston, was in Tascosa at the same time and outshot Bat Masterson, sports writer and United States Marshall, while Billy won in a money match. The Kid excelled in most sports of that day, but did not drink. No doubt his men drank enough for him and then some.[17]

One evening, Dr. Hoyt and Billy, who had become good friends, left a dance that was being held in the Pedro Romero home and strolled into the town plaza. They decided to run a race, and although Billy was fast, Hoyt outdistanced him. As they neared the Romero home, the doctor slowed down but Billy ran through the door, tripping and falling and sliding along the floor to the center of the room. In seconds, his pals, thinking he was being attacked, surrounded him, guns drawn. They were barred from other dances in Tascosa from then on.[18]

Dr. Hoyt encouraged Billy to go to Mexico or some distant place and begin life anew. He gave Billy a watch which the young outlaw admired, and when the doctor left Tascosa Billy gave him a handsome Arabian race horse named Dandy Dick. As a joke he wrote out the following bill of sale for the animal:[19]

Tascosa, Texas
Thurs. Oct. 25, 1878.

Know all persons by these presents, that I do hereby sell and deliver to Henry F. Hoyt, one sorrel horse branded B.B. on left hip and other indistinct brands on shoulders, for the sum of seventy-five dollars ($75.00 in hand received.)

W. H. Bonney.
Witnesses, James E. McMasters
George W. Howard.

Dr. Hoyt later sent his friend, Charles A. Siringo at Carrizozo, New Mexico, a photostatic copy of the bill of sale. Siringo showed the paper to his friend, James Brady, court reporter there. When Brady read the description of the horse and saw Billy's signature on the paper, he said, "My God, it was my father's horse. Murphy formerly owned the animal and gave it to my father when he became sheriff." [20]

Disposing of the 125 stolen horses in Texas, Billy and his friends began to steal cattle from the High Plains ranchmen. They contracted with Pat Coughlin of White Oaks, New Mexico, for the sale of the cattle. At one time they stole 116 head from Chisum. Jim Carlyle, leader of a posse sent to capture the Kid, was killed in 1880. Chisum heard that Billy was selling stolen cattle to Coughlin so he led an armed group of men to Coughlin's slaughter pen in Fort Stanton but found no proof of stolen jinglebob cattle.[21]

To put a stop to Billy's rustling, Panhandle cattlemen met in Mobeetie, and organized the Panhandle Cattle Raisers Association, making Charles Goodnight president.[22] They then sent an expedition to scout for Texas cattle driven into New Mexico by Billy and his gang. They rode to White Oaks where the cattle were said to have been sold. Although they found many hides with Texas brands on the corral fence they could find no evidence of those who had sold the cattle.

Pat Garrett was now the deputy sheriff of Lincoln county and the cowmen urged him to hunt down Billy. Garrett and Billy had once been friends but Garrett put this friendship aside when he became an officer of the law. During his term of deputy, Sheriff Garrett searched the country east of Fort Sumner for Billy but found no trace. Billy heard about the search and sent the following letter to Governor Wallace:

> . . . Deputy Sheriff Garrett acting under Chisum's orders went to Portales and found nothing. J.S. Chisum is the man who got me into trouble and has benefitted thousands by it, and is now doing all he can against me.[23]

In his book *Tragic Days of Billy The Kid*, Frazier Hunt tells about a talk made by John Meadows at the Old-Timers Reunion in Roswell in 1931 that shed light on the relationship between Pat Garrett, the newly elected sheriff, and Billy the Kid. Meadows said that Garrett had told him the same story time and time again. Garrett told him that his first object was to settle the trouble between him and Billy without bloodshed. "I figured how I could do it," he said, and continued:

> Before I was elected sheriff I saw the Kid and talked with him, played poker with him and told him the best thing he could do was to get up and be gone three or four years, and then come back and there would be nothing said about it. There would be no trouble. If you fellows don't do this, and stay here, I am going to arrest you or kill you, or you are going to kill me. But Billy would not see the point.[24]

Garrett wrote that Billy looked for support from John Chisum after the death of Tunstall and McSween since they were the three for whom he had "worked, fought and killed." Billy thought Chisum should support him but Chisum refused to do so and his negative attitude caused Billy to hate him and to swear vengeance against him. As a result Billy and his companions began to drive off Chisum's cattle. At one time Billy stole 188 head and, pretending to be working for Chisum, sold them to Colorado beef buyers for $600. Chisum traced the herd and drove the cattle back to his range, but Billy was $600 to the good.[25]

Billy and his gang became more wary after Garrett was elected sheriff. They were careful about appearing in public. Only Billy's close companions remained with him, including Tom O'Folliard, Charlie Bowdre, Dave Rudabaugh, Tom Pickett and Billy Wilson. Some of Billy's gang had left him in Tascosa. Among those were Henry Brown, Fred Waite and John Middleton, who gave up the life and went straight. Brown became a peace officer in Tascosa. Middleton and Waite went to Indian Territory. They urged Billy to quit the life also but he refused.[26]

Sheriff Garrett and his deputies continued to search for Billy and his gang. They eventually ran into them on a cold winter night near Fort Sumner. O'Folliard was shot in the ensuing gun battle. Billy and his remaining men found refuge in a small rock house on the Taiban River. Bowdre was shot and killed as he walked to the door carrying a feed bag. Garrett's posse heard Billy say, "Charlie, you're done for. Go and see if you can get one of them before you die." After hours of arguing between Garrett and the Kid, the latter agreed to surrender since the cabin was bitterly cold and the men had no food. The remaining outlaws, Billy, Rudabaugh, Pickett and Wilson then surrendered to Garrett.[27]

Billy and his companions were taken to Las Vegas by the sheriff's posse, and later to Santa Fe, where Billy was tried for the murder of Sheriff Brady.[28] Pickett was later freed but Rudabaugh was convicted of mail robbery and murder and thrown in jail. He later escaped.[29] It is not known what happened to Wilson. He was probably released with Pickett.

The jury found the Kid guilty and Judge Bristol sentenced him to be hanged at Lincoln between nine and three o'clock conforming to the New Mexico law that put execution dates thirty days after sentence. Billy accepted the verdict calmly. He did not denounce the judge or the court, although there are several legends to this effect. Evidently Billy still had hope that Governor Wallace would grant him executive clemency.[30]

Taken back to Lincoln, Billy was locked up in the office of Sheriff Garrett on the second floor of the two-story courthouse since Lincoln had no jail. Robert Ollinger and J. W. Bell were hired to guard him.[31]

Guarded night and day, Billy had little chance of escaping his jailers, but he was constantly watching for an opportunity. On April 28, Ollinger left Bell in charge of Billy and went out to eat. Bell picked up a newspaper and was glancing over it when Billy slipped his small hands out of the handcuffs and hit Bell over the head with them. Grabbing Bell's pistol he coldly murdered him, then kicked open the door and grabbed Ollinger's shotgun. Standing by the window, he waited for Ollinger to return. When he saw him crossing the dusty street he called "Hello Bob," and when Ollinger looked up in shocked surprise, "eighteen buckshot hit him in the breast."[32]

Many citizens watched this cold-blooded drama and

some felt sympathy for Billy the Kid. They heard Billy call down to a boy in the street to bring him Sheriff Garrett's horse. The lad hurried off and was soon back with the county clerk's pony instead, saddled and ready to go. Jumping into the saddle Billy dashed down Main Street, but turned before leaving town to wave his hat and shout, "Three cheers for Billy the Kid." [33]

After escaping from jail in April, 1881, Billy rendezvoused with friends and avoided Garrett and his deputies for several months. Heavy awards were posted for his capture in public places.

The editor of the *Laredo Times*, wrote in the August 10, 1881, issue of his paper that there seemed to be but one man in all that section with nerve enough to follow Billy, and that man was Sheriff Garrett who was on his trail and, unless killed himself, would get the Kid eventually. The editor predicted also that Chisum would soon start some of his own cowboys after the Kid, and they in conjunction with Sheriff Garrett would surely run him down before too long.[34]

Sheriff Garrett and two of his deputies, John W. Poe and Thomas K. McKinney, continued to hunt for Billy. One morning a Texan whom Poe had befriended brought him a message. He said the Kid was in the house of his sweetheart at Fort Sumner. Poe carried the message to Sheriff Garrett in Lincoln. They rode to Roswell for Special Deputy McKinney and then continued on to Fort Sumner. Billy was not found where they expected him to be, and it was thought wise that Sheriff Garrett talk with Pete Maxwell, well-known ranchman in the area who was living at that time in one of the government buildings.[35]

Garrett found Maxwell in bed as it was about midnight. He asked him if he knew where Billy was staying. Maxwell had seen the Kid around but did not know whether he was still in the community or not.

There was a Mexican's adobe cabin back of Maxwell's home where Billy was hiding. He was reading a newspaper upon the bed, his boots on the floor. He soon threw down the paper and ordered the Mexican to cook him a steak. He was told there was no meat in the house. "I'll see if Pete has any," Billy said, and he ran across the yard barefooted with a butcher knife in his hand.[36] Poe and McKinney remained in the shadows at the end of the porch while Garrett went inside. As they whispered in the dark, Poe noticed a man walking inside the picket fence in front of the building. The man was within three or four feet of them before he saw them. *Quien es? Quien es?* (Who is it?) he cried in Spanish as he backed through the door Garrett had entered.[37] Backing into Pete's room Billy asked, "Who is out there Pete?" Receiving no answer he backed away, asking again "Who in the h--l is in there?"

Pete whispered to Garrett, "That's him Pat," Garrett's pistol fired and Billy slumped to the floor shot through the heart.

"I killed the Kid," Pat cried. "I killed the Kid." [38]

Thus ended the life of William H. Bonney, one of the coolest-headed and most daring young outlaws that ever lived," wrote Siringo in his book, *A Texas Cowboy.*

"Perhaps other bad men rode across the stage as the pioneer years rolled by and might have stolen more cattle but no one made open warfare in the bold, audacious manner of the Kid and his men," wrote McCarty.[39]

Rafaelita Pryor wrote, "Regardless of how people of today regard the Kid, to a lot of his contemporaries he was a hero." [40]

Billy the Kid was born in New York, November 23, 1859, the son of William H. and Kathleen Bonney.[41] Billy's father died when he was a small lad and his mother moved west and married a man named Antrim in Santa Fe, New Mexico, March 1, 1873. Her son, a witness to the wedding, was recorded as Henry McCarty, the name he was known by in Silver City where the family settled. Later he was called Billy Antrim after his stepfather. He did not use the name Bonney until he reached Pecos Valley. In 1875 when working in Graham County, Arizona, his friends began calling him "Billy the Kid," or "Kid," due to his youthful appearance. While working as a teamster in Graham County in 1877, he killed his first man. A blacksmith, Frank Cahill, nicknamed "Windy" due to his overbearing manner, delighted in teasing the Kid. One day, Billy had enough and, snatching the blacksmith's pistol, shot Cahill, who died the next day. Billy was thrown in jail, but escaped into old Mexico.

It was through the influence of Jesse Evans that Billy rode into Lincoln County, New Mexico in 1877.[42]

After Governor Lew Wallace came to New Mexico as the new governor, he had a conference with Billy the Kid — a peace talk it might best be called. Billy told the governor some facts about several of the men who played a prominent part in the Lincoln County War. As a result Governor Wallace later wrote Billy a letter telling him if he would surrender and be tried in court, stating the facts he had told the governor about participants in the war, his life would be spared and he would be freed.[43]

Billy agreed to do so at first, but later refused, afraid it was a ruse to capture him. It was said that he told Wallace, "I've gone too far now. The only justice I could get in this country would be six-shooter justice." [44]

Dr. Hoyt later met Governor Wallace who told him if Billy had kept the appointment his life would have been saved and he might have become a useful citizen.[45]

Mullin wrote in *The Boyhood of Billy the Kid*:

> Only when that factional conflict called the Lincoln County War was ended, when Tunstall and McSween had been killed and their supporters had become fugitives, did the partisan fighting man become the common horse-thief, the cattle-rustler, the desperado, which made the name of Billy the Kid notorious.[46]

Billy the Kid is buried in Fort Sumner, New Mexico. A strong iron fence protects his tombstone which was badly chipped away before the fence was erected. This author visited the grave in 1940.

In looks and personality Pat E. Garrett was the personification of one's imaginary picture of the pioneer sheriff who killed Billy the Kid. Six feet–four inches tall, mild mannered, soft-spoken, nattily dressed, with a "killer" mustache, his tall, slim figure was a familiar one on the streets of many southwestern pioneer towns. He became famous after taking the life of Billy the Kid in July 1881. Garrett was an Alabamian by birth, born June 5, 1850, but spent most of his boyhood in Louisiana. He drifted into Texas when nineteen years of age and worked as a cowboy for several years on a Dallas County ranch. He was in Fort Griffin, Texas, during buffalo hunting days and no doubt took part in several Indian raids.

Garrett married Polinaria Gutierrez in Fort Sumner, New Mexico in 1880 and fathered seven children. He first met Billy the Kid in that settlement and they became friends. It is thought that Billy told him the story of his early life that was later recorded in Garrett's book, *The Authentic Life of Billy the Kid.*

Garrett served as captain of a Texas Ranger Company in the Panhandle of Texas and later operated a horse ranch near Uvalde where he was a friend of John Nance Garner. He served as Sheriff of Doña Ana county in New Mexico, was a hunting companion of Emerson Hough and other notables, and was a friend of President Teddy Roosevelt. The latter appointed him customs collector at El Paso for four years.

Garrett was shot and killed en route to Las Cruces, New Mexico, from his ranch in the Organ Mountains February 29, 1908. Wayne Brazil, a tenant on his ranch admitted the killing, was tried in court and found innocent. Many questions remain unanswered about Garrett's death. It was commonly believed that he was the victim of a murder plot.[47]

Chapter 16

John Chisum Dies and The Jinglebob Ranch Falls

When John Chisum built his baronial South Spring Ranch home, he designed a long hall, or dogtrot through the center and directly over a gurgling stream. Here a cooling shade could be found morning or afternoon on the hottest day. Failing in health, Chisum was no longer the vigorous ranchman who had met and usually conquered adversity so many times over the years. He had never gotten back on his feet financially after the Lincoln County War and his herds were greatly diminished. It will be recalled that he sold some 30,000 head to Hunter, Evans and Company and turned over $27,000 of his best cattle to his brother, Pittser. He had repaid an old debt to Coggin and Wylie with 8,000 jinglebob cattle.[1]

Sallie called the Lincoln County War "a big jolt" to her Uncle John, which took a big toll. His ranching was, of course, neglected during the big cattle feud and rustlers took advantage of the opportunity to increase their herds with mavericks from the jinglebob range. Drouth struck the country in the 1880s, and soon after the nester cowman moved in and claimed much of the range that Chisum had once considered his own special empire.[2]

In 1882 he had sold the west end of his ranch to the Littlefield Cattle Company, fully a twenty-mile front on the Pecos including the Bosque Grande Ranch.[3]

Despite poor health and advancing years Chisum still had a vital interest in upgrading his remaining herd, and he was still considered one of the leading cowmen in the Southwest in the early eighties. The *Denver Daily News*, April 11, 1883, praised his cattle:

> One of the finest bunches of cattle in New Mexico is controlled by John S. Chisum — his herd numbers 30,000 head of the best graded cattle —last year 6000 calves were branded with the U on the shoulder, the distinguished mark of the cattle king.

The article stated that jinglebob cattle were always in demand and found steady markets. Several thousand were sold in May, 1884 at Las Vegas, New Mexico, and a large herd was delivered to Dodge City for Medicine Lodge parties.

A growth had appeared on Chisum's neck in 1883.[4] He was concerned. Excepting a case of smallpox which he survived in 1877, he had never been sick in his life. Reports state that he consulted local doctors who operated on the growth. Could one of those doctors have been Dr. Hoyt, his old friend in Tascosa? Chisum did not improve after the operation and, although ill and lacking in strength, attended cattlemen's meetings and represented Lincoln County at the New Mexico Cattle Raisers Association in Santa Fe. He was a member of the Lincoln County Stock Association and was paid $1,000 by this organization in the spring of 1884 for recovery of cattle stolen from his range.[5]

Chisum also made plans to locate a trainload of cattle on an Arizona ranch along the San Pedro River. He had occupied that range in 1870. This news soon spread over the cow country of Arizona and rubbed Arizona ranchmen the wrong way. They did not want Chisum's jinglebob cattle to dominate the range in their territory as those herds had done in New Mexico. An article in the *Yuma Arizona Star*, September 20, was sympathetic to Chisum:

> It is difficult to conceive why Mr. Chisum has not an equal right

> with any other American citizen to buy land and graze cattle — there is something ludicrous in the idea — that intelligent men will interfere with the vested legal rights of John Chisum or any other man.[6]

Cattlemen of Arizona stopped worrying about jinglebob cattle taking over the range when they learned of Chisum's failing health.

Chisum underwent surgery in Kansas City, Missouri July 7, 1884.[7] While convalescing he was cheered to learn that he had been appointed commissioner to represent the cattle-raising industry in the territories at the Southern Industrial Exposition in New Orleans, December 1, through the spring of 1884. He was unable to do so, due to his poor health.[8]

Chisum's health steadily worsened. His saddle hung empty in the barn and his favorite mount grazed on the range. As Evetts Haley wrote in his book *George Littlefield, Texan*, "Chisum sat in the retreating shade on the west side of the dog-run in the morning and in the lengthening shadows on the west side of a hot afternoon, his lanky frame draped comfortably over a chair, awaiting death to set him free from an incurable disease." [9]

In desperation, and with little hope, Chisum took the advice of specialists and left home again in late July for Eureka Springs, Arkansas, hoping the baths would help him. Saying a sad farewell to his family, Chisum left his South Spring Ranch home, no doubt knowing in his own heart that he would never return. The baths seemed to bring some relief but he continued to fail and death came December 22, 1884.[10]

Christmas festivities were interrupted on the ranch when a lone rider arrived at the headquarters gate, his horse bathed in sweat, to tell of Chisum's death. The news was soon relayed to all cow camps, towns and plazas. A *baile* was in progress at Fort Sumner when the news was heard. The caller stopped the dancing and broke the news. "John Chisum is dead. We have lost one of New Mexico's most popular citizens." [11]

Evetts Haley, probably described Chisum better than any other writer:

> Cut to no ordinary pattern, Chisum stood out alone as an unusual man in an unusual and distinctive land, not alone in the length of his Long Rail brand, the oddity of his jinglebob earmark and the breadth of his extensive range, but also in his personality and character which seem to baffle analysis. A brave man, involved in the Lincoln County War, the bloodiest southwestern saga ever sung by 30-30 bullets and .45 slugs.[12]

Chisum's brother James had joined him in Eureka Springs before he died, and John requested him to take him back to Paris to be buried so he could sleep beside his parents in the old family cemetery. His wish was carried out. Interment was held on Christmas Day, 1884.[13]

The Golden Era, Lincoln, January 1, 1885, reported Chisum's death:

> Our readers will be pained to learn of the death at Eureka Springs, Ark. of John S. Chisum, which occurred at that place about two weeks ago. Mr. Chisum was one of the pioneers and frontiersmen of Lincoln County, having come here at a very early day and has been identified with the history ever since. Eccentric in many ways, gruff in manner, yet he was always a warm friend and no man ever looked closer after the pleasure and comfort of the men under his employ during his long experience as a cattleman.

One writer wrote that Chisum was a soldier although he wore no uniform and his only weapons against blood-thirsty Indians and cattle thieves "were a cool, clear business head and a desire to be a square shooter with his fellowman."[14]

Chisum's famous Jinglebob Ranch began to go downhill after his death. His brothers, Pittser and James, could not hold the ranch together. Pittser withdrew from the company, took his share of stock, married and moved back to Texas.[15] The remaining stockholders, James and his two sons, Walter and William, and his daughter, Sallie, and her husband, William Robert, held the ranch together for a few years, calling it The Jinglebob Land and Cattle Company. The company closed in 1891.[16]

Sallie Chisum, once "The Queen of the Jinglebob," spent

her last days in Roswell, New Mexico. It was told that Negro Frank, faithful as always, and loving the Chisum heirs as his own family, financed Sallie in buying a ranch after she and her husband separated.

Many cowboys worked for John Chisum during his long years of ranching, but no doubt the most loyal was Negro Frank. Chisum bought Frank when he was about four years old and he had remained with his old master all his life, an eye witness to the exciting scenes and tragedies that occurred in the Pecos Valley during pioneer days. In his older years he was a respected citizen in Roswell and a source of authentic Chisum history. He had a remarkable memory.[17] There was no better line rider nor horse wrangler on the plains than Negro Frank. He had been reared in the saddle and had been Chisum's right-hand man ever since he was old enough to ride a horse.

Frank accumulated his own small herd through the years. When the Chisum heirs merged their stock in a new company, Negro Frank refused to include his cattle. He had his individual VF brand and said, "If I turn my cattle into the company, you would not know who they belonged to, but when you see a cow branded VF you can say it is one of Negro Frank's cows." Negro Frank was lucky. When the company folded, he had his herd and it had increased to several hundred head.[18]

Negro Frank lived in Roswell several years after the death of John Chisum. He had a dream one night that caused him to go back to his Texas home. He said his old master had sent him a message, saying "We want you." As a result, Frank left Roswell, telling his friends he was going back home to die and to be buried near his master, John Chisum. Negro Frank Chisum, who took his master's name, is still remembered as a colorful pioneer character along the Pecos.[19]

Frank Collinson summed up the chapter he wrote about Chisum as follows:

> Today if you should ask a man along the Pecos where the old Bosque Grande Ranch of John Chisum used to be, he would

look at you and say that he never heard of Bosque Grande, nor of Chisum.[20]

This writer does not agree with Collinson. As long as the Old West lives (and it will never die), stories of the Jinglebob King of the Pecos, trailblazer and pioneer of the Southwest, "who rode to fortune and power in his shirt sleeves on a cow pony;" the Lincoln County War, and Billy the Kid, will be told and retold as long as the treacherous Pecos threads through the wild canyons of the West.

* * * * * *

In April 1884, an omnibus suit was filed against John Chisum, his brothers, Pittser and James, and William Robert, son-in-law of James, in the district court of Lincoln County by Catron and Thornton and other prominent Santa Fe attorneys.

Ten plaintiffs were represented in the suit, all asking the same relief. Plaintiffs charged Chisum had transferred all his cattle, horses and personal property to Hunter, Evans and Company to prevent plaintiffs and other creditors from recovering their indebtedness.

The complaint alleged that Hunter, Evans and Company would possess Chisum's property until all claims were settled — that in 1880 Chisum had directed the firm to convey the remaining property, including large herds of cattle, to his brothers Pittser and James. The brothers had no personal property, the suit went on, and did not pay one dollar to Hunter, Evans and Company for the transfer in their behalf. Further, the suit claimed the brothers and William Robert, son-in-law of James, had confederated together in furtherance of a sacred trust, at all times but tools and employees in the hands of John Chisum.[21]

John Chisum was suffering with terminal cancer when the suit was filed and was unable to appear in court in his own behalf. The trial judge entered a decree November 14, 1884, in favor of the plaintiffs. It was the judge's opinion that the transfer of property by Chisum had been prejudicial to his creditors. F.L. Vanderveer was appointed to sell $57,030.86 worth of Chisum property to pay the plaintiffs. John Chisum never knew the court's verdict. He died eight months after the suit was filed, prior to its conclusion.[22]

Sallie Chisum Robert, Chisum's favorite, and very pretty niece. His cowboys called her "The Queen of the Jinglebob."

Meady Chisum and her husband, Bob Jones. Meady was the daughter of John Chisum and a slave girl.

Meady Chisum and husband Bob Jones at old Roanoke, Texas home.

Meady Chisum Jones in front of the old Jones home at Roanoke, Texas. Meady was the daughter of John Simpson Chisum.

The old Chisum home near Denton, Texas

Jinks Jones, son of Meady and Bob (John Dalford) Jones and a grandson of John Chisum, who operated an auction barn near Grapevine, Texas until his death.

Eugie Thomas feeding her chickens.
This picture was made several years ago.

Eugie Thomas, ninety-nine years of age. The last living grandchild of John Simpson Chisum.

Billy the Kid "had no shoulders, small feet and hands, long blond hair and gray-blue eyes that never stopped looking around."

— Courtesy *The Cattleman*

ENDNOTES

Chapter 1: The Famous Mark of John Simpson Chisum

1. Clarke, Mary Whatley, "Cowboy Collinson," *The Cattleman*, August 1957.

2. Potter, Colonel Jack, "The Jingle-bob Herd," *New Mexico*, July 1945.

3. Wellman, Paul I., *The Thundering Herd* (New York, Doubleday and Co. 1951) 34–35.

4. Potter, Colonel Jack, "The Jingle-bob Herd," *New Mexico*, July 1945.

5. Potter. [See Note 4.]

6. Adams, Clarence S. and Tom E. Brown, *Three Ranches West: A Story of John S. Chisum* (New York: Carlton Press, 1972, 85).

7. Adams. [See Note 4.]

8. Haley, J. Evetts, *Charles Goodnight, Cowman and Plainsman*, (Norman, University of Oklahoma Press, 1949), 235.

9. Hinton, Harwood P. Jr. "John Simpson Chisum, 1877–84" (Master of Arts Thesis, Columbia University, 1955). Later published in *New Mexico Historical Review* (NMHR), XXXI (July 1956), 177–205, (October 1956), 310–37; (January 1957), 53–65.

10. Hinton, "Chisum," *NMHR* XXXI, 192.

11. Clarke, Mary Whatley, "Cowboy Collinson," *The Cattleman*, August 1957.

12. Clarke. [See Note 11.]

13. Twitchell, Ralph Emerson, *Leading Facts of New Mexico History*. Vol. III, 297.

14. Funk and Wagnalls Standard Reference Encyclopedia (New York: Standard Reference Works Pub. Co., Inc., 1959–1960, Vol. 7), page 2512.

15. Potter, Colonel Jack, "Ruidoso Ghost Steer of the Pecos," *The Cattleman*, June 1930.

16. Potter. [See Note 15.]

Chapter 2: John Simpson Chisum Begins his Ranching Saga

1. Ridings, Sam P., *The Chisholm Trail* (Guthrie, Oklahoma; Co-operative Pub. Co., 1936) Chapter IV, 42.

2. Taylor T.U., "John Simpson Chisum," *Frontier Times*, 1932. (Republished June–July, 1978.

3. Taylor. [See Note 2.]

4. Collinson, Frank, *Life in the Saddle*, Edited and arranged by Mary Whatley Clarke. (Norman, Oklahoma) University of Oklahoma Press, 1963, page 137.

5. Cross, Cora Milton, "Stories of Old Trail Drivers of Long Ago."

Gus Gildea one-time Texas Cowboy, Trail Driver, Ranger and Sheriff, Tells of Cattle Rustling and Comanche and Apache Raids. (Dallas, Texas; *Semi-Weekly Farm News*, March 20, 1930. Gildea, A.M. (Gus), "Experiences of a Ranger and Scout," *The Trail Drivers of Texas* (Nashville, Tenn.: Cokesbury Press, 1925). Compiled and edited by J. Marvin Hunter. Published under direction of George W. Saunders, page 975.

6. Hinton, "Chisum" *NMHR* XXXI, 180.

7. Collinson, Frank, *Life in the Saddle*, 137.

8. Hinton, "Chisum" *NMHR* XXXI, July, 1957, page 183.

9. Fulton, Maurice G. *History of the Lincoln County War* (Tucson, Arizona: The University of Arizona Press, 1968) Edited by Robert M. Mullin, page 33.

10. Cox, James. *Historical and Biographical Record of the Cattle Industry and the Cattlemen of Texas and Adjacent Territory, 1895* (St. Louis; Tiernan Printing Co., 1895) 299–302 (Hereafter cited as Cox, *Cattle Industry 1895)*.

11. Webb, Walter Prescott, Editor-in-Chief, *The Handbook of Texas*. H. Bailey Carroll, managing editor. (Austin; The State Historical Association 1952, Vol. I), 342. (Hereafter cited as *Handbook of Texas*.)

12. Cox, *Cattle Industry 1895* pages 299–302.

13. Taylor, T.U. "John Simpson Chisum," *Frontier Times*, June–July, 1978.

14. Douglas, C.L. *Cattle Kings of Texas*, Fort Worth, Branch–Smith Inc. 1968, page 113.

15. The Handbook of Texas, Vol. I, p. 342: Cox, *Cattle Industry* 1895, pages 299–302.

16. ——— [See Note 15.]

17. ——— [See Note 15.]

18. Bridges, C.A. *History of Denton, Texas From Its Beginning to 1960.* Texian Press, 1970, Waco, Texas, pages 20–21–24.

19. ——— [See Note 18.]

20. ——— [See Note 18.]

21. Taylor, T.U. "John Simpson Chisum," *Frontier Times*, June–July, 1978.

22. Bridges, C.A. *History of Denton, Texas From Its Beginning to 1960.* Texian Press, Waco, 1978, page 23

23. Taylor, T.U. "John Simpson Chisum," *Frontier Times*, June–July, 1978.

24. ——— [See Note 23.]

25. Gard, Wayne, *Waco-Tribune Herald*, October 30, 1939.

26. Hinton, "Chisum," *NMHR* XXXI, 186.

27. ——— [See Note 26.]

28. Cox, *Cattle Industry 1895,* pages 299–302.

29. ——— [See Note 28.]

30. Told to the author by Mrs. Eugie Jones Thomas, at her Grapevine

home. She was then ninety-two years of age. Facts also given by Mrs. Jinks Jones, Grapevine, Texas.

Chapter 3: Chisum Moves Herds to Concho River

1. Taylor, T. U. "John Simpson Chisum." *Frontier Times*, June–July 1978.
2. Cox, *Cattle Industry 1895*, pages 299–302.
3. Hinton, "Chisum," *NMHR*, XXXI, 187.
4. Taylor, T. U. "Trailing John Chisum to New Mexico," *Frontier Times*, June, 1936.
5. *Handbook of Texas*, Vol. II, page 800.
6. Thompson, Albert W., "Old-Time Cattle Outfits in Pecos Valley," *The Cattleman*, September 1943.
7. Thompson. [See Note 6.]
8. Thompson. [See Note 6.]
9. Cox, *Cattle Industry 1895*, pages 299–302.
10. *Texas Almanac*, 1969, page 86.
11. Cox, *Cattle Industry 1895*, pages 299–302.
12. Cox. [See Note 11.]
13. *The Handbook of Texas*, Vol. I, page 832.
14. Collinson, Frank, *Life in the Saddle*, page 139.
15. Collinson, [See Note 14.] page 140.
16. Haley, J. Evetts, *Charles Goodnight, Cowman and Plainsman*, University of Oklahoma Press, Norman, 1949, page 259.
17. ———. "Fort Concho and the Texas Frontier." *San Angelo Standard Times*, 1952, page 37 (Reconnaissances, Lieutenant French).
18. Taylor, T. U., "Trailing John Chisum to New Mexico," *Frontier Times*, June, 1936.
19. Taylor. [See Note 18.]
20. Taylor. [See Note 18.]
21. Taylor. [See Note 18.]
22. Redfield, Georgia B., "The Historic Chisum Trail," *The Cattleman*, 1945.

Chapter 4: Chisum Makes a Deal With Goodnight

1. Cox, *Cattle Industry 1895*, pages 209–302.
2. Cox. [See Note 1.]
3. Cox. [See Note 1.]
4. Wallis, George A., "Cattle Kings," *New Mexico*, June 1930.
5. Cox, *Cattle Industry 1895*, pages 209–302.
6. Frank Warner Angel's Notes on New Mexico Territory, 1878, edited by Lee Scott Theison, *Arizona and the West*, Vol. 18, Number four, Winter, 1978.

7. Fulton, *History of the Lincoln County War,* page 35.

8. Cox, *Cattle Industry 1895.* pages 299–302.

9. Clarke, Mary Whatley, *The Palo Pinto Story,* The Manney Co. Fort Worth, 1955, 62–63.

10. *The Handbook of Texas,* Vol. II, page 86, Vol. I, page 709.

11. ———. [See Note 10.] Vol. I, p. 709

12. Holland, G. A., *History of Parker County and the Double Log Cabin,* The Herald Pub. Co. Weatherford, Texas, 1937, page 81.

13. Holland. [See Note 12.]

14. Cox, *Cattle Industry 1895,* pages 299–302.

15. Cox. [See Note 14.]

16. Haley, Evetts, *Charles Goodnight, Cowman and Plainsman,* page 232.

17. Goodnight, Charles, "More About the Chisholm Trail," *The Trail Drivers of Texas,* page 950, edited by J. Marvin Hunter, Cokesbury Press Nashville, Tennessee, 1925. Published under direction of George W. Saunders, President of the Old Time Trail Drivers Association.

18. Hinton, "Chisum," *NMHR,* XXXI 190.

19. Cox, *Cattle Industry 1895,* pages 299–302.

20. Fulton. [See Note 7.] page 34.

21. Cox, *Cattle Industry 1895,* pages 299–302.

22. Keleher, William A., *The Fabulous Frontier,* Chapter 2, The Rudal Press, Santa Fe, New Mexico, 1945.

Chapter 5: John Chisum Makes Two Major Decisions

1. Hinton, "Chisum" *NMHR,* XXXI, July 1956. 189.

2. Hinton. [See Note 1.]

3. McLean, William Hunter, *From Ayr to Thurber; Three Hunter Brothers and the Winning of the West.* Fort Worth Genealogical Society, 1978, 23. (Hereafter McLean.)

4. McLean. [See Note 3.] page 18–19.

5. McLean. [See Note 3.] page 119.

6. McLean. [See Note 3.] page 119.

7. McLean. [See Note 3.] page 72.

8. Final Decree of Court in Chancery at Hillsboro, New Mexico, November 14, 1835. Deed Book A., 457–65, Chaves County, New Mexico.

9. Cox, *Cattle Industry, 1895,* 299–302.

10. Cox. [See Note 9.]

11. Cox. [See Note 9.]

12. Hinton, "Chisum" *NMHR* XXXI, 191–192, July 1956.

13. McCarty, John K. *Maverick Town, The Story of Old Tascosa,* Norman; University of Oklahoma Press, 1948, 76.

14. Wellman, Paul I. *The Trampling Herd.* (New York; Doubleday and Co. 1951) 335–336.

15. Frank W. Angel's Report 1878, *Arizona and the West*, vol. 18 Number IV. Winter 1978.
16. Fulton, 33.
17. McLean. [See Note 3.] page 72.
18. McLean. [See Note 58.] page 58.
19. McLean. [See Note 3.]
20. McLean. [See Note 3.] pages 75–81.
21. McLean. [See Note 3.] page 80.
22. McLean. [See Note 3.]
23. McLean. [See Note 3.] page 81.
24. McLean. [See Note 3.]

Chapter 6: "Judge Lynch" Takes Over

1. Gildea, A. M. (Gus) "Experiences of a Ranger and Scout," *The Trail Drivers of Texas*, 975–986; Cora Milton Cross, "Stories of the Trail Drivers of Long ago, Gus Gilea One-Time Texas Cowboy, Trail Driver, Ranger and Sheriff, Tells of Cattle Rustling and Comanches and Apaches raids." *Semi-Weekly Farm News,* Dallas, Texas, March 20, 1930; George A. Wallis, *Cattle Kings of the Staked Plains* American Guild Press, Dallas.
2. Collinson, Frank, *Life in the Saddle*, 141–142.
3. Collinson. [See Note 2.]
4. Collinson. [See Note 2.]
5. Hinton, "Chisum" *NMHR*, XXXI, 182.
6. Keleher, William A. *The Fabulous Frontier*. The Royal Press, Santa Fe, 1945, Chapter 2 about John S. Chisum.
7. Gildea, A.M. (Gus), "Experiences of a Ranger and Scout," *The Trail Drivers of Texas*, 975–986.

*Note: Where did Gus, a simple New World cowboy, hear the expression "Bakesheesh," or "Bakshish," which for hundreds of years has rung familiarly in streets and alleys of Middle Eastern cities, called out by beggars enlisting alms from passerbys. Doubtless the word "Valese" is corrupted Spanish, deriving from Vayase, the Spanish word for Go, in the command sense, i.e., Be off with you.

8. Gildea. [See Note 7.]
9. Gildea. [See Note 7.]
10. Gildea. [See Note 7.]
11. Hinton, "Chisum" *NMHR* XXXI, 183; also Dane Coolidge *Fighting Men of the West*, 50.
12. Keleher, William A. *The Fabulous Frontier, Twelve New Mexico Items*, Rudal Press, Santa Fe, 1945, 52.
13. Dobie, Frank, *Cow People*, Little Brown and Co. Boston, 1954, 201–262.
14. Hinton, "Chisum," *NMHR*, XXXI, 183.
15. Halye, J. Evetts, *Charles Goodnight Cowman and Plainsman*, 235.

16. ——— *George W. Littlefield Texan.* University of Oklahoma Press, Norman, 1945, 183.

17. *Prose and Poetry of the Livestock Industry of the United States*, Denver, National Livestock and Historical Association 1904, 502–3.

18. Collinson, Frank, *Life in the Saddle*, 145.

19. Brothers, Mary Hudson, *A Pecos Pioneer.* The University of New Mexico Press, Albuquerque, 1943, 49.

Chapter 7: The Chisum War on the Pecos

1. *Santa Fe New Mexican,* Feb. 8, 1876; Fulton, pages 35–36.

2. Fulton, page 36.

3. Hinton, "Chisum" *NMHR*, XXXI, 197.

4. Hinton. [See Note 3.], page 198.

5. Fulton, page 37.

6. Hinton, "Chisum" *NMHR* XXXI, 199–200.

7. *Mesilla Independent,* June 15, 1877.

8. Fulton. [See Note 1.], page 50.

9. Fulton. [See Note 1.], page 37.

10. Fulton. [See Note 1.], page 40.

11. Fulton. [See Note 1.], page 41.

12. Keleher, *Fabulous frontier*, page 52.

13. Klasner Lily *My Girlhood Among Outlaws*, University of Arizona Press, Tucson, 1972. Edited by Eve Bell, pages 258–259.

Chapter 8: Causes of the Lincoln County War

1. McCarty, John K. *Maverick Town, The Story of Old Tascosa*, University of Oklahoma Press, Norman, 1948. 76–77.

2. Cox, *Cattle Industry*, 1895, 299–302.

3. *Dictionary American Biography*, Charles Scribner's Sons, New York 1930, Vol. IV.

4. By permission from Maurice G. Fulton's *History of the Lincoln County War*, Robert N. Mullin, editor, Tucson; University of Arizona Press copyright 1968.

5. Anderson, George B. *History of New Mexico,* 223–228.

6. Wallace, Lew. *An Autobiography*, Harper and Row Publishers, Inc. New York, 1906, 914–915.

7. Fulton, Maurice G. *History of the Lincoln County War*, The University of Arizona Press, Tucson, Arizona, 51.

8. McLean, William Hunter, *From Ayr to Thurber*, 83–84.

9. Keleher, *Violence in Lincoln County*, Foreword, p. XIV.

10. Pryer, Rafaelita, "Siege of the McSween House," *Frontier Times* April–May 1969.

11. Hough, Emerson, *The Story of the Cowboy*, D. Appleton and Co., 1898, New York, 302–303.

12. Anderson, George, *History of New Mexico*, Vol. 2, 1023–1025.

13. Fulton, [See Note 7.] Caption under Chisum's photograph, page 32.

14. Clarke, Mary Whatley, "Cowboy Collinson," *The Cattleman*, August 1957.

15. Pryer, Rafaelita, "Siege of the McSween House," *Frontier Times*, April–May, 1969.

16. Hinton, "Chisum" *NMHR*, XXXI, 193.

17. Keleher, *Fabulous Frontier*, 184–249.

18. Fulton, [See Note 7.] page 52.

19. Keleher, *Violence in Lincoln County*, 55.

20. Fulton, 54–58.

21. Keleher, *Violence in Lincoln County*, 52.

22. Cunningham, Eugene, "Fought with Billy the Kid," *Frontier Times*, March, 1932.

23. Fulton, 95.

24. Cunningham, Eugene, "Fought With Billy the Kid," *Frontier Times*, March, 1932.

Chapter 9: The Killing of Tunstall and Sheriff Brady

1. Hinton, "Chisum" *NMHR*, XXXI, 207–203.

2. Fulton, Maurice Carland, *History of the Lincoln County War*, 66–67.

3. Fulton. [See Note 2.] 83–84.

4. Garrett, Patrick Floyd, *The Authentic Life of Billy the Kid*, Norman, University of Oklahoma Press, 1954, new edition, 48.

5. Cunningham, Eugene, "Fought With Billy the Kid," *Frontier Times*, March 1932.

6. Nolan, Frederick W. *John Henry Tunstall*, University of New Mexico Press, Albuquerque, New Mexico, 1965. 364.

7. "Frank Warner Angel's Notes on New Mexico Territory 1878," *Arizona and the West*, Vol. 18, Number 4.

8. ———. [See Note 7.]

9. Hunt, Frazier, *The Tragic Days of Billy the Kid*, 39.

10. Clarke, Mary Whatley, "Cowboy Collinson" *The Cattleman*, Aug. 1957.

11. Hunt, Frazier, *The Tragic Days of Billy the Kid*, 35.

12. Clarke, Mary Whatley, "Cowboy Collinson," *The Cattleman*, August 1937.

13. Garrett, Patrick Floud, *The Authentic Life of Billy the Kid*, 57.

14. Cunningham, Eugene, "Fought With Billy the Kid," *Frontier Times*, March 1932; George Anderson, *History of New Mexico*, 225; Fulton 175.

15. Diringo, Charles A. *History of Billy the Kid*, Facsimile reproduction Steck-Vaughn Co. Publishers, Austin, Texas, 1967, 49–52.
16. Hinton, "Chisum" *NMHR*, XXXI, 317.
17. Hinton. [See Note 16.] 315.
18. Hinton. [See Note 16.] Footnote 19, page 325, October 1956.
19. Coleman, Max W. "J.K. Millwee, Frontiersman," *Frontier Times*, February 1937, Vol. 14.
20. Hinton, "Chisum" *NMHR*, XXXI, 323.

Chapter 10: The Arrests of Chisum and McSween

1. Wellman, Paul I, *The Trampling Herd*, pages 332–336.
2. Fulton. [See Note ——, Chapter II.], page 51.
3. Fulton. [See Note ——, Chapter ——.], page 73.
4. Keleher, *Violence in Lincoln County*, page 59.
5. Hinton, "Chisum" *NMHR*, XXXI, 194.
6. Hinton. [See Note 5.], page 311; Angel Report, 1878, page 24.
7. Hinton. [See Note 5.]
8. Hinton. [See Note 5.]
9. Keleher, *Violence in Lincoln County*, pages 42–43.
10. Keleher. [See Note 9.]
11. Keleher. [See Note 9.], pages 48–49.
12. Haley, Evetts, *Charles Goodnight, Cowman, Plainsman*, pages 233–35.
13. Fulton, [See Note ——, Chapter ——.], pages 96–97.
14. Keleher, *Violence in Lincoln County*, page 47.
15. Keleher. [See Note 14.], page 45.
16. Keleher. [See Note 14.], pages 44–47.
17. Hinton, "Chisum" *NMHR*, XXXI, 187.
18. Hinton. [See Note 17.], page 188.
19. Keleher, *Violence in Lincoln County*, pages 45–46.
20. Keleher. [See Note 19.], pages 46–47.
21. Keleher. [See Note 19.], page 47.
22. Fulton, page 35.
23. Hinton, "Chisum" *NMHR*, XXXI, 314.
24. Hinton. [See Note 23.], page 311.
25. Haley, Evetts, *Charles Goodnight, Cowman, Plainsman*, page 335.
26. Wellman, *The Trampling Herd*, page 335.
27. Haley, Evetts, *Charles Goodnight, Cowman, Plainsman*, page 234.
28. Wellman, *The Trampling Herd*, page 334.
29. Wellman. [See Note 28.], page 335.
30. Siringo, Charles, *Riata and Spurs*, page 56.
31. McLean, *From Ayr to Thurber*, page 84.
32. Potter, Colonel Jack, "The Jinglebob Herd," *New Mexico*, July 1945.

Chapter 11: Spring Term of Court Keeps Grand Jury Busy

1. Fulton, *History of the Lincoln County War,* page 201.
2. Fulton. [See Note 1.].
3. Fulton. [See Note 1.], page 203
4. Fulton. [See Note 1.]
5. Fulton. [See Note 1.]
6. Fulton. [See Note 1.], page 204
7. Fulton. [See Note 1.]
8. Hinton, "Chisum," *NMHR*, XXXI, July 1956, 320.
9. Hinton. [See Note 8.], page 320.
10. Hinton. [See Note 8.], page 321.
11. Keleher, *Fabulous Frontier*, page 54.
12. Fulton. [See Note 1.], pages 34–35.
13. Letter to author, August 7, 1978.
14. Final Decree of Court in Chancery at Hillsboro, New Mexico, November 14, 1885.
15. Klasner, Lily, *My Girlhood Among Outlaws*, page 254.
16. Haley, Evetts, *Charles Goodnight*, page 234.
17. ———, Charles Goodnight, page 234.

Chapter 12: The Siege of the McSween Home and His Death

1. Keleher, *Violence in Lincoln County*, page 141.
2. Fulton, pages 121–127.
3. Keleher, *Violence in Lincoln County*, page 267.
4. Keleher. [See Note 3.], page 269.
5. Keleher. [See Note 3.], page 140.
6. Fulton. [See Note 1.], page 135.
7. Keleher, *Violence in Lincoln County*, page 141.
8. Keleher. [See Note 7.], page 142.
9. Keleher. [See Note 7.], page 142.
10. Keleher. [See Note 7.], pages 142–143.
11. Keleher.[See Note 7.], pages 143–144.
12. Anderson, George, *History of New Mexico*, page 226.
13. Fulton, pages 265–266.
14. Hunt, Frazier, *The Tragic Days of Billy The Kid.* Hastings House Publishers, New York, 1956, page 91.
15. Siringo, Charles, *A Texas Cowboy*, William Sloane Asso., New York, 1950, page 170.
16. Hunt, Frazier, *The Tragic Days of Billy The Kid.* Hastings House Publishers, New York, 1956, page 95.
17. Fulton. [See Note 2.], pages 265–266.
18. Keleher, *Violence in Lincoln County*, page 144.

19. Pryor, Rafaelita, "Siege of the McSween House," *Frontier Times*, April–May, 1969.

20. Pryor. [See Note 19.]

21. Fulton. [See Note 2.], page 269.

22. Fulton. [See Note 2.]

23. Mary Whatley (Clarke), "Cowboy Collinson," *The Cattleman*, February 1958.

24. Fulton. [See Note 2.], page 270.

25. Fulton. [See Note 2.], pages 273–274.

26. Fulton. [See Note 2.], page 275.

27. Fulton. [See Note 2.], page 270; Nolan, John Henry Tunstall, page 381 (similar letter).

28. Keleher, *Violence in Lincoln County*, page 192.

29. Keleher. [See Note 28.], page 213.

30. Fulton. [See Note 2.], page 364.

31. Keleher, *Violence in Lincoln County*, page 149.

32. Keleher. [See Note 31.]

33. Fulton. [See Note 2.], pages 419–421.

34. Fulton. [See Note 2.], page 421.

35. Fulton. [See Note 2.], pages 268–289.

36. Potter, Colonel Jack, *The Cattleman*, March 1947.

37. Wellman, Paul I., *The Trampling Herd*, page 351.

38. Mclean, *From Ayr to Thurber*, page 83.

39. McLean. [See Note 38.], page 87.

40. McLean. [See Note 38.], page 88.

41. McLean. [See Note 38.], page 86.

42. McLean. [See Note 38.], page 85.

43. McLean. [See Note 38.], page 86.

44. McLean. [See Note 38.], page 87.

45. McLean. [See Note 38.], page 88.

46. McLean. [See Note 38.], pages 42–43–27–28–32–33.

Chapter 13: Removal to Texas; Governor Wallace Comes to New Mexico

1. Fulton, *History of the Lincoln County War*, page 289.
2. *Independent*, May 11, 1879.
3. Poe, *Buckboard Days*, page 162.
4. *Mesilla News*, August 10, 1878.
5. McCarty, *Maverick Town, The Story of Old Tascosa*, page 41.
6. Hinton, "Chisum," *NMHR*, XXXI, 327.
7. Hinton. [See Note 6.]
8. McCarty, *Maverick Town, The Story of Old Tascosa*, page 41
9. *Handbook of Texas*, Vol. II, page 172.

10. Clarke, Mary Whatley, "Cowboy Collinson," *The Cattleman*, August 1957.

11. Clarke. [See Note 10.]

12. Clarke. [See Note 10.]

13. Clarke. [See Note 10.]

14. Clarke. [See Note 10.]

15. Collinson, Frank. *Life in The Saddle*, University of Oklahoma Press, Norman (edited and arranged by Mary Whatley Clarke), page 121.

16. Clarke, Mary Whatley, "Cowboy Collinson," *The Cattleman*, August 1957.

17. Cox, *Cattle Industry 1895*, 299–302.

18. Cox. [See Note 17.]

19. Keleher, *the Fabulous Frontier*, 52–53.

20. Fulton. [See Note 1.], 169.

21. Fulton. [See Note 1.], 170.

22. Fulton. [See Note 1.]

23. Fulton. [See Note 1.], 171.

24. Funk and Wagnalls Standard Reference Encyclopedia, Standard Reference Works Pub. Co. Inc. New York, 1959–1960, 9059.

25. Wallace, Lew, *An Autography*, New York; Harper and Brothers 1906, Vol. II, 914–915.

26. Hinton, "Chisum" *NMHR*, XXXI, 329–330.

27. Hinton. [See Note 26.], 330.

28. Keleher, *Violence in Lincoln County*, 225.

29. Hinton, "Chisum" *NMHR*, XXXII, 60.

30. Haley, J. Evetts, *George Littlefield Texan* 137–41.

Chapter 14: South Spring River Ranch — An Oasis in the Desert

1. Redfield, Georgia B. "Queen of the Jingle-Bob," *The Cattleman*, December 1945.

2. Keleher, *Violence in Lincoln County*, Note, page 59.

3. Poe, Sophia A. *Buckboard Days*, The Caxton Printers Ltd. Caldwell, Idaho. 1936. page 165.

4. Poe. [See Note 3.], page 16.

5. Hinton, "Chisum," *NMHR*, XXXII, 59.

6. Redfield, Georgia B. "Queen of the Jingle-Bob," *The Cattleman*, December 1945.

7. Hinton, "Chisum," *NMHR*, XXXI, 334 T.C. Richardson, *East Texas; Its History and Its Makers*, New York. Lewis Historical Pub. Co. 1940, Vol. III, page 1109.

8. Brothers, Mary Hudson, *A Pecos Pioneer*, The University of New Mexico Press, 1943, page 51.

9. Collinson, Frank, *Life in the Saddle*, page 138.

10. Hinton, "Chisum," *NMHR*, XXXI, 334.

11. Hinton. [See Note 10.]

12. Redfield, Georgia B. [See Note 6.], "Queen of the Jingle-Bob," *The Cattleman*, December 1945.

13. Anderson, George, *History of New Mexico*, Vol. II, page 796.

14. Hinton, "Chisum," *NMHR*, XXXI, 334.

15. Redfield, Georgia B. "Queen of the Jingle-Bob," *The Cattleman*, December 1945.

16. Potter, Jack. "The Jingle-Bob Herd" *New Mexico*, 1945.

17. Potter. [See Note 16.]

18. Potter. [See Note 16.]

19. Potter. [See Note 16.]

20. Redfield, Georgia B. "Queen of the Jingle-Bob" *The Cattleman*, December 1945.

21. Poe, Sophie A. *Buckboard Days*, page 158.

22. Poe. [See Note 21.], page 165.

23. Hinton, "Chisum," *NMHR*, XXXI, 181.

24. Hinton. [See Note 23.], page 180.

25. Klasner, Lily, *My Girlhood Among Outlaws*, page 5.

26. Klasner. [See Note 25.], page 71.

27. Klasner. [See Note 25.], pages 240–241.

28. Interview with Emory Odell Jones and Mrs. Jinks Jones.

Chapter 15: Billy The Kid

1. Hunt, Frazier, *The Tragic Days of Billy the Kid*, Hastings House Publishers, New York, 1950, 117.

2. Siringo, Charles A. *History of Billy the Kid*, 81.

3. Cunningham, Eugene, "Fought With Billy the Kid," *Frontier Timers*, March 1932.

4. McLean, *From Ayr to Thurber*, page 86.

5. Clarke, Mary Whatley, "Cowboy Collinson" *The Cattleman*, August 1957.

6. Clarke. [See Note 5.]

7. Clarke. [See Note 5.]

8. Siringo, Charles A., *History of Billy the Kid*, 52.

9. Garrett, Pat, *The Authentic Life of Billy the Kid*, 89.

10. Anderson, George B., *History of New Mexico*, 227.

11. Fulton, Maurice Garland, *History of the Lincoln County War* 379–380.

12. Fulton. [See Note 11.], 380.

13. McCarty, *Maverick Town the Story of Old Tascosa*, 81.

14. McCarty. [See Note 13.], 49.

15. Hoyt, Henry P. *A Frontier Doctor*, Boston, Houghton-Mifflin Co. 1929, 50–51.

16. Hoyt. [See Note 15.], 92.

17. McCarty, *Maverick Town the Story of Old Tascosa*, 79.

18. McCarty. [See Note 17.], 79–80.
19. McLean, *From Ayr to Thurber*, 89.
20. McCarty, Maverick Town, the Story of Old Tascosa, 81.
21. Hinton, "Chisum" *NMHR*, XXXII, 58; C. L. Sonnichsen, *Tularosa Last of the Western West*. The Eevin-Adair Co. N. H. 1960, 233.
22. Clarke, Mary Whatley, *A Century of Cow Business*, Evans Press, Fort Worth, Texas, 1970, 130–131.
23. Wallace Collection, letter dated Dec. 12, 1880.
24. Hunt, Frazier, *The Tragic Days of Billy the Kid*, 214.
25. Garrett, Pat, *The Authentic Life of Billy the Kid*, 32–37.
26. McCarty, *Maverick Town the Story of Old Tascosa*, 81.
27. McCarty. [See Note 26.]
28. McCarty. [See Note 26.], 88.
29. Cunningham, Eugene, "Fought With Billy the Kid," *Frontier Times*, March, 1932.
30. Fulton, Maurice Garland, *History of the Lincoln County War*, 390.
31. Cunningham, Eugene, "Fought With Billy the Kid," *Frontier Times*, March 1932.
32. Siringo, *History of Billy the Kid*, (Facsimile reproduction) Steck-Vaughn Co. Austin, 1967, 15–17.
33. Siringo. [See Note 32.]
34. *Laredo Times*, August 10, 1882, editorial comment at end of story.
35. McCarty, *Maverick Town the Story of Old Tascosa*, 91.
36. McCarty. [See Note 35.]
37. McCarty. [See Note 35.], 92.
38. Garrett, Pat, *The Authentic Life of Billy the Kid*, 144–145.
39. McCarty, *Maverick Town the Story of Old Tascosa*, 91.
40. Pryor, Rafaelita, "The Siege of the McSween House," *Frontier Times*, April–May 1969.
41. Siringo, Chas. A., *History of Billy the Kid*, page 41.
42. Fulton, Maurice Garland, *History of the Lincoln County War*, 67–68.
43. Hoyt, Henry F. *A Frontier Doctor*, 118.
44. Cunningham, Eugene, "Fought With Billy the Kid," *Frontier Times*, March 1942.
45. Hoyt, Henry F., *A Frontier Doctor*, page 118.
46. Mullin, Robert W., *The Boyhood of Billy the Kid. Texas Western Press*, El Paso, 1967, (Southern studies) Vol. V, No I, Samuel D. Myres, Editor).
47. Garrett, Pat, *The Authentic Life of Billy the Kid.* Introduction, J. C. Dykes: William A. Keleher, "Patrict Floyd Garrett," *Twelve New Mexico Items*. The University of New Mexico Press, Albuquerque, 1962.

Chapter 16: John Chisum Dies and The Jinglebob Ranch Falls

1. Potter, Colonel Jack. "Fall of the Jingle-Bob Empire," *The Cattleman*, March 1947.

2. Potter. [See Note 1.]
3. Potter. [See Note 1.]
4. Hinton, "Chisum" *NMHR* XXXI, 63–64.
5. Hinton. [See Note 4.], 62.
6. Hinton. [See Note 4.], 64.
7. Potter, Col. Jack, "Fall of the Jingle-Bob Empire," *The Cattleman*, March 1947.
8. Hinton, "Chisum" *NMHR*, XXXI, 65.
9. Haley, J. Evetts, *George Littlefield Texan*, page 148.
10. Potter, Colonel Jack, "Fall of the Jingle-Bob Empire," *The Cattleman*, March 1947.
11. Potter. [See Note 10.]
12. Haley, J. Evetts, *George Littlefield Texan*, page 148.
13. Hinton, "Chisum" *NMHR* XXXI, 65.
14. Redfield, Georgia B. "The Queen of the Jingle-Bobs," *The Cattleman*, November 1945.
15. Hinton, "Chisum" *NMHR* XXXI 62.
16. Fulton, 411.
17. Anderson, George, *History of New Mexico*, Vol. 2, page 1023.
18. Potter, Colonel Jack, "Fall of the Jingle-Bob Empire," The Cattleman, March, 1947.
19. Potter. [See Note 18.]
20. Collinson, Life In The Saddle, 145.
21. Keleher, *Violence in Lincoln County*, 50.
22. Keleher. [See Note 21.], 49–51.

Bibliography

BOOKS

Adams, Clarence S. and Tom E. Brown. *Three Ranches West.* A true story of John S. Chisum. Carlton Press, Inc., New York, N.Y., 1972.

Anderson, George B. *History of New Mexico.* 2 vols. Los Angeles, Pacific States Publishing Company, 1907.

Bancroft, Robert Howe. *History of Arizona and New Mexico, 1530–1888.* San Francisco. The History Publishers 1889 (An Arno Press book published in cooperation with McGraw–Hill Co., N.Y., Toronto, London, Sydney.

Bates, Ed. F. *History and Reminiscences of Denton County.* Terrill Wheeler Printing, Inc., Denton, Texas, 1976. (Formerly McNitzky Printing Co). Special Bicentennial Reprint. First printing 1918 by McNitzky Printing Co., Denton.

Bridges, C. A. *History of Denton, Texas From Its Beginning to 1960.* Texian Press, Waco, 1978.

Brothers, Mary Hudson. *A Pecos Pioneer.* Albuquerque, New Mexico, University of New Mexico Press, 1943.

Burns, Walter Noble. *The Saga of Billy the Kid.* Garden City, New York: Doubleday–Page and Co., 1926.

———. *Tombstone An Iliad of the Southwest.* Grosset and Dunlap, Pub., New York, 1929.

Coan, Charles F. *A History of New Mexico.* 3 vols. Chicago and New York, 1932.

Coe, George W. *Frontier Fighter.* Albuquerque: University of New Mexico Press, 1951.

Collinson, Frank. *Life in the Saddle.* Edited and arranged by Mary Whatley Clarke. Norman, University of Oklahoma Press, 1963.

Coolidge, Dane. *Fighting Men of the West.* Dutton and Co., Inc., New York, 1932.

Cox, James. *Historical and Biographical Record of the Cattle Industry and the Cattlemen of Texas and Adjacent Territory 1895.* St. Louis, Woodward and Tiernan Printing Co., 1895.

Dictionary of American Biography, Charles Scribner's Sons, New York, 1930, Vol. 4.

Dobie, J. Frank. *A Vaquero of the Brush Country*, Dallas, 1929.

———. *Coronado's Children*, Garden City Publishing Co., Garden City, N.Y., 1930.

———. *Cow People.* Little Brown and Co., Boston, 1964.

———. *The Voice of the Coyote.* Lincoln, University of Nebraska Press, 1961.

———. *The Longhorns.* Bramhall House, N.Y., 1940.

Douglas, C. L. *Cattle Kings of Texas*, Fort Worth, Branch–Smith, Inc., 1968.

Erwin, Allen A., *The Southwest of John Horton Slaughter*. Glendale, Cal., The Arthur H. Clark Co., 1965.

Fullen, Louise and Artrip. *Memories of Daniel Fore (Jim) Chisholm and the Chisholm Trail*. Boobeville, Ark., 1949.

Fulton, Maurice S. *History of the Lincoln County War*. Tucson, Arizona. Edited by Robert M. Mullin, University of Arizona Press, 1968.

Gard, Wayne. *The Chisholm Trail*. Norman, University of Oklahoma Press, 1954.

Garrett, Pat F. *The Authentic Life of Billy the Kid*. New York, McMillan Co., 1927. Edited by Maurice Fulton. New edition copyright, Norman, University of Oklahoma Press, 1954.

Haley, J. Evetts. "Fort Concho and the Texas Frontier." *San Angelo Standard Times*, San Angelo, Texas, 1952.

———. *Charles Goodnight, Cowman and Plainsman*. Norman, University of Oklahoma Press, 1949.

———. *George Littlefield, Texan*. Norman, University of Oklahoma Press, 1943.

———. *The XIT Ranch of Texas*. Chicago, Illinois, The Lakeside Press, 1929. Revised edition, University of Oklahoma Press, 1963.

Halsell, H. H. *Cowboys and Cattleland*. Wilkinson Printing Co., Dallas, 1945.

Hamlin, William Lee. *The True Story of Billy the Kid*. The Caxton Printers Ltd., Caldwell, Idaho, 1959.

Hendricks, George D. *The Bad Men of the West*. San Antonio, The Naylor Co., 1942.

Hinkle, James F. *Early Days of a Cowboy on the Pecos*. Stagecoach Press, Santa Fe, 1965.

Holden, William Curry. *A Ranching Saga. The Lives of William Electious Halsell and Ewing Halsell*. Trinity University Press, San Antonio, 1976.

Hollon, W. Eugene. *The Southwest Old and New*. Lincoln, University of Nebraska Press, 1961.

Horgan, Paul. *Great River, The Rio Grande in North American History*. Holt Rinehart and Winston, New York, Chicago, San Francisco, Vol. I, 1965.

Horn, Calvin. *New Mexico's Troubled Years*. Horn and Wallace Publishers, 1963, Albuquerque.

Hough, Emerson. *North of 36*. D. Appleton, New York, 1923.

———. *The Story of the Cowboy*. Gosset and Dunlap, New York, 1905, p. 208.

Hoyt, Henry F., *A Frontier Doctor*. Boston, Houghton-Mifflin Co., 1929.

Hunt, Frazier. *The Tragic Days of Billy the Kid*, Hastings House Publishers, N.Y. 1956.

Hyer, Julian. *The Land of Beginning Again*. Tupper and Love Inc. Atlanta, 1952.

Keleher, William A. *Fabulous Frontier*. Santa Fe, The Rydal Press, 1945. Revised edition, University of New Mexico Press, 1962.

———. *Violence in Lincoln County 1869–1881*. Albuquerque, New Mexico, University of New Mexico Press, 1957.

———. *Turmoil in New Mexico 1846–1968*. The Rydal Press, Santa Fe, New Mexico, 1952.

Kendall, George Wilkins. *Kendall's Narrative of the Santa Fe Expedition*, Vol. III. Austin, The Steck Co., 1935.

Klasner, Lily, *My Girlhood Among Outlaws*. Tucson, Arizona, The University of Arizona Press, 1972.

Lea, Tom. *The King Ranch*. Boston, Massachusetts, Little, Brown and Co., 1957.

Linn, John J. *Reminiscences of Fifty Years in Texas*. New York, D. and J. Sadler and Co., 1882.

Lord, John. *Frontier Dust*, Hartford, Connecticut, 1926.

McCarty, John K. *Maverick Town. The Story of Old Tascosa*. Norman, University of Oklahoma Press, 1948.

McKee, Irving. *"Ben-Hur" Wallace, the Life of General Lew Wallace*. Berkeley, California, University of California Press, 1947.

McLean, William Hunter. *From Ayr to Thurber, Three Hunter Brothers and the Winning of the West*. Fort Worth Genealogical Society. New Printing Co., 1978.

Mullin, Robert N. *The Boyhood of Billy The Kid*. Texas Western Press, El Paso, 1967. (Southern Studies, Vol. V, No. 1, Samuel D. Myres, Editor.)

New Mexico, American Guide Series. Hastings House, New York, 1940.

Nolan, Frederick W. *John Henry Tunstall*. Albuquerque, New Mexico, University of New Mexico Press, 1965.

Nordyke, Lewis. *Cattle Empire*. New York, W. Murrow, 1949.

Pearce, W.N. *The Matador Land and Cattle Company*. Norman, University of Oklahoma Press, 1964.

Poe, Sophie A. *Buckboard Days*. Edited by Eugene Cunningham, Caxton Printers, Caldwell, Idaho, 1936.

Potter, Jack M. *Lead Steer and Other Tales*. Leader Press, Clayton, New Mexico, 1939.

Prose and Poetry of the Livestock Industry of the United States. Kansas City, 1915. National Livestock Historical Association.

Raine, William Macleod. *Famous Sheriffs and Western Outlaws*. Garden City Publishing Company, Garden City, N.Y., 1929.

Richardson, T.C. *East Texas—Its History and Its Makers*. New York, Lewis Historical Publishing Company, 1940, Vol. III, p. 1109.

Richardson, Rupert Norval. *Texas The Lone Star State*. New York, Prentice–Hall, 1943.

Ridings, Sam P. *The Chisholm Trail*. Guthrie, Oklahoma, Cooperative Publishing Company, 1936.

Scobee, Barry. *Old Fort Davis*. San Antonio, Texas, The Naylor Company, 1947.

Siringo, Charles A. *A Texas Cowboy*. William Sloan Associates, New York, 1950. (First printing Dobson Publishing Company, Chicago, 1888).

———. *Riata and Spurs*. Houghton–Mifflin Company, Boston and New York, 1931.

———. *History of Billy the Kid*. Steck–Vaughn Company Publishers, Austin, Texas, 1967. (A facsimile reproduction)

Smith, Sidney W. *From the Cow Camp to the Pulpit*. Cincinnati, Ohio, The Christian Lead Corporation, 1927.

Sonnichsen, C. L. *Tularosa Last of the Western West*. The Devin–Adair Company, New York, 1960.

Texas Almanac and State Industrial Guide, 1968–1969. A. H. Belo Corporation, *The Dallas Morning News*, 1968–1969.

The Handbook of Texas. Walter Prescott Webb, Editor-in-Chief. H. Bailey Carroll, managing editor. The Texas State Historical Association 1952, 2 vols.

The Trail Drivers of Texas. Compiled and edited by J. Marvin Hunter. Published under the direction of George W. Saunders, Cokesbury Press, Nashville, Tennessee, 1925.

Thrall, Homer S. *A History of Texas From Earliest Settlements to the Year 1877*. University Publishing Company, New York, 1876.

Twitchell, Ralph Emerson. *Leading Facts of New Mexico History*. Cedar Rapids, Iowa, The Torch Press, 1911–1917, 5 vols.

Wallace, Lew. *An Autobiography*. Harper and Brothers Pub. Co. New York, 1906.

Wallis, George A. *Cattle Kings of the Staked Plains*. American Guild Press, Dallas, Texas, 1957.

Walters, Lorenzo D. *Tombstone's Yesterday*. Tucson, Ache Printing Co., 1928. Reprint publication, Glorieta, N.M., Rio Grande Press, 1968.

Webb, Walter Prescott, *The Great Plains*. Boston: Ginn and Co., 1931.

———. *The Texas Rangers*. Boston, Houghton–Mifflin Co., 1955.

Wellman, Paul I. *The Trampling Herd*. New York: Doubleday and Co., 1951.

White, Owen P. *Them Was the Days; From El Paso to Prohibition*. New York, Minton, Balch and Co., 1925.

Winfrey, Dorman H. *Indian Papers, 1825–1843*. Austin, Texas State Library, 1959.

———. *Texas Indian Papers, 1844–1845*. Texas State Library, 1960.

———. *Texas Indian Papers, 1846–1859*. Texas State Library, Austin, 1960.

Wortham, Louis J. *A History of Texas from Wilderness to Commonwealth*. 5 vols. Vol 5. 135–177. Wortham–Molyneaux Co., Fort Worth, 1924.

Yoakum H. *History of Texas From Its First Settlement in 1685 to Its Annexation to the United States in 1846*. Redfield, Beekman Street, New York, 1855. A facsimile reproduction in one volume, Steck–Vaughn Co., Austin, Texas

Zuber, William P. *My Eighty Years in Texas.* Austin, The University of Texas Press, 1971, Edited by Janis Boyle Mayfield.

PRINTED ARTICLES

Ashton, John. "Texas Cattle Trade in 1870." *The Cattleman.* July, 1951.

"Chisum vs Chisholm," *Frontier Times,* Vol. 7, No. 9, June 1929. (Letter to J. Marvin Hunter from Alex Williams, about correct spelling of name.)

Clarke, Mary Whatley, "Cowboy Collinson," *The Cattleman,* August 1957.

———. "Cowboy Collinson," *The Cattleman,* Jan. 1958.

———. "Cowboy Collinson," *The Cattleman,* Feb. 1958.

Coleman, Max M. "J.K. Milliwee, Frontiersman," *Frontier Times,* Aug., 1928.

Cunningham, Eugene, "Fought With Billy the Kid," *Frontier Times,* March 1932.

Dobie, J. Frank, "Billy the Kid." *Southwest Review,* Vol. XIV, No. 3, Spring 1929.

Douglas, C.L., "The Jingle-Bob King," *The Cattleman,* Feb. 1969.

Erwin, Helen, "When Billy the Kid was Brought to Trial." *Frontier Times,* March 1929; *Star Telegram,* Fort Worth, Dec. 2, 1928.

Hendron, J.W. "No Pardon for Billy," *Frontier Times,* March, 1951.

Hinton, Harwood P., Jr. "John Simpson Chisum, 1877–84," *New Mexico Historical Review,* Vol. XXXI, No. 3, July, 1956.

———. "John Simpson Chisum 1877–84," *New Mexico Historical Review,* Vol. XXXI, No. 4, October 1956.

———. "John Simpson Chisum 1877–84," *New Mexico Historical Review,* Vol. XXXII, No. I, January 1957.

Holt, R.D. "Old-Time Cowmen of Coleman County," *The Cattleman,* Sept. 1933.

Hubert, Harry, "Old Trickham of Cattle Trail Days," *Frontier Times,* Vol. 27, No. 5, 1950.

Hunter, J. Marvin, Sr. "Pat Garrett Slayer of Billy the Kid." *Frontier Times,* Vol. 27, No. 5, 1950.

Keleher, W.A. "Maxwell Land Grant" (A New Mexico Item), *Argosy Antiquarian Ltd.,* New York, 1942.

Kelly, Jean. "Early Texas Cattle Baron Recalled," By granddaughter, *The Cattleman,* April 1969.

Moses, Tad. "Early Day Cattleman," *The Cattleman,* November 1947.

Mullins, James W. "Texas Trail Is Discussed," *The Cattleman,* July 1926.

"Pioneer Men Tell of Early Days in Coleman County," *Frontier Times,* Oct. 1949.

Potter, Col. Jack. "The Jingle-Bob Herd," *New Mexico Magazine,* July 1945.

———. "Fall of the Jingle-Bob Empire," *The Cattleman,* Mar. 1947.

———. "Ruidoso Ghost Steer of the Pecos," *The Cattleman,* June 1930.

———. "Fort Sumner on the Pecos," *The Cattleman*, Feb. 1950.

Pryor, Rafaelita, "Siege of the McSween House," *Frontier Times*, April-May 1969.

Record, Ralph H. "At the End of the Texas Trail, Range Riding and Ranching," *West Texas Historical Asso. Yearbook*, XIX, Oct. 1943.

Redfield, Georgia B. "The Historic Chisum Trail," *Frontier Times*, Dec. 1945.

———. "Queen of the Jingle-Bob Empire," *The Cattleman,* Dec. 1945.

Robertson, Sug. "Prose and Poetry of the Livestock Industry of the United States," *National Livestock Historical Assn. Yearbook,* 1904.

Saunders, George W. "More Lore of Cattle Trails," *The Cattleman*, Sept. 1926.

Sinclair, John L. "On the Hoof." *New Mexico Magazine*, October 1939.

Sutton, Fred E. "Dr. Hoyt, Panhandle's First Doctor," *Frontier Times*, Aug. 1930.

Taylor, T. U. "Trailing John Chisum to New Mexico," *Frontier Times*, June 1936.

———. "John Simpson Chisum and His Kin," *Frontier Times*, Dec. 1937.

———. "Life Sketch of John Simpson Chisum," *Frontier Times*, May 1936.

———. "John Simpson Chisum," *Frontier Times*, June-July, 1978.

Thompson, Albert W. "Old Time Cattle Outfits in Pecos Valley," *New Mexico*, Dec. 1943.

Wallis, George A. "Cattle Kings," *New Mexico*, June 1936.

NEWSPAPERS

Fort Worth Press, October 6, 1963.

Grant County Herald, June 8, 1877.

Grant County Herald, Dec. 28, 1878.

Kansas City Livestock Indicator, July 1, 1878.

Kansas City Livestock Indicator, Mar. 7, 1889.

Laredo Times, Aug. 10, 1881.

Las Vegas Gazette, June 15, 1878.

Las Vegas Gazette, Aug. 16, 1878.

Las Vegas Gazette, Aug. 17, 1878.

Las Vegas Gazette, Aug. 6, 1880.

Las Vegas Gazette, Nov. 10, 1880.

Mesilla News, July 25, 1874.

Mesilla News, Jun. 10–25, 1876.

Mesilla News, Aug. 10, 1879.

Mesilla Valley Independent, June 23, 1877.

Mesilla Valley Independent, July 21, 1877.

Mesilla Valley Independent, Oct. 6, 1877.

Mesilla Valley Independent, April 13, 1878.

Mesilla Valley Independent, April 18, 1878.

Mesilla Valley Independent, May 11, 1879.

Mesilla Valley Independent, May 24, 1879.
New Mexican, May 4, 1878.
Old Abe Eagle, Lincoln, Sept. 1, 1892.
Roswell Record, Oct. 7, 1937.
Semi-Weekly Farm News, Dallas, Mar. 10, 1930.
Semi-Weekly Farm News, Dallas, Feb. 1, 1954.
The Dodge City Globe, April 6, 1876.
The Clarksville Standard (Texas), Aug. 14, 1852.

MISCELLANEOUS

Deed Book G., p. 540, Lamar County, Paris Texas (Chisum land) Abstract No. 278 General Land office, Austin, Texas, Pre-emption Certificate No. 156, 3–16–59. Patent No. 39, Vol. 24, Fannin 3rd Class, File No. 2390.

Letter from Mrs. McSween to John P. Tunstall. *Anarchy in Lincoln County*, p. 181.

The Story of John Simpson Chisum as told by Eugie Jones Thomas to Yvonne A. Jenkins, Unpublished, Denton, Texas Library.

Captain Henry W. Strong. "My Frontier Days and Indian Fights on the Plains of Texas." Unpublished, Denton, Texas Library.

Interviews with Mesdames Jinks Jones and Eugie Jones Thomas, Sept. 9, 1977.

Angel, Frank W. Special Agent 1878. File No. 44–48. Record Group 60. Department of Justice Records, National Archives, Washington, D.C. Report on death of John H. Tunstall.

INDEX